Beatrice Harraden

Ships That Pass in the Night

Beatrice Harraden

Ships That Pass in the Night

ISBN/EAN: 9783337812508

Printed in Europe, USA, Canada, Australia, Japan

Cover: Foto ©berggeist007 / pixelio.de

More available books at **www.hansebooks.com**

SHIPS THAT PASS
IN THE NIGHT

BY
BEATRICE HARRADEN

ELEVENTH EDITION

LONDON
LAWRENCE & BULLEN
16 HENRIETTA STREET, COVENT GARDEN, W.C.
1893

TO MY DEAR FRIENDS,

AGNES AND JOHN KENDALL,

THIS LITTLE BOOK,

WRITTEN MOSTLY IN THEIR HOME,

IS LOVINGLY DEDICATED.

Jan. 12th, 1893.

CONTENTS.

PART I.

CHAPTER		PAGE
I.	A New-Comer	1
II.	Which contains a Few Details	5
III.	Mrs. Reffold learns her Lesson	10
IV.	Concerning Wärli and Marie	20
V.	The Disagreeable Man	25
VI.	The Traveller and the Temple of Knowledge	35
VII.	Bernardine	41
VIII.	The Story moves on at last	52
IX.	Bernardine preaches	60
X.	The Disagreeable Man is seen in a New Light	70
XI.	"If One has made the One Great Sacrifice"	93
XII.	The Disagreeable Man makes a Loan	104
XIII.	A Domestic Scene	120
XIV.	Concerning the Caretakers	135
XV.	Which contains Nothing	144
XVI.	When the Soul knows its own Remorse	158
XVII.	A Return to Old Pastures	161
VIII.	A Betrothal	182
XIX.	Ships that speak each other in passing	188
XX.	A Love-letter	196

PART II.

I.	The Dusting of the Books	205
II.	Bernardine begins her Book	216
III.	Failure and Success: A Prologue	218
V.	The Disagreeable Man gives up his Freedom	222
V.	The Building of the Bridge	231

SHIPS THAT PASS IN THE NIGHT.

PART I.

CHAPTER I.

A NEW-COMER.

"YES, indeed," remarked one of the guests at the English table, "yes, indeed, we start life thinking that we shall build a great cathedral, a crowning glory to architecture, and we end by contriving a mud hut."

"I am glad you think so well of human nature," said the Disagreeable Man, suddenly looking up from the newspaper which he always read during meal-time. "I should be more inclined to say that we end by being content to dig a hole, and get into it, like the earth men."

A silence followed these words; the English community at that end of the table was struck with astonishment at hearing the Disagreeable Man speak. The few sentences he had spoken during the last four years at Petershof were on record; this was decidedly the longest of them all.

"He is going to speak again," whispered beautiful Mrs. Reffold to her neighbour.

The Disagreeable Man once more looked up from his newspaper.

"Please, pass me the Yorkshire relish," he said in his rough way to a girl sitting next to him.

The spell was broken, and the conversation started afresh. But the girl who had passed the Yorkshire relish sat silent and listless, her food untouched, and her wine untasted. She was small and thin; her face looked haggard. She was a new-comer, and had, indeed, arrived at Petershof only two hours before the *table-d'hôte* bell rang. But there did not seem to be any nervous shrinking in her manner, nor any shyness at having to face the two hundred and fifty guests of the Kurhaus. She seemed rather to be unaware of their presence; or, if aware of, certainly indifferent to the scrutiny under which she was being placed. She was recalled to reality by the voice of the Disagreeable Man. She did not hear what he said, but she mechanically stretched out her hand and passed him the mustard-pot.

"Is that what you asked for?" she said half dreamily; "or was it the water-bottle?"

"You are rather deaf, I should think," said the Disagreeable Man placidly. "I only remarked that it was a pity you were not eating your dinner. Perhaps the scrutiny of the two hundred and fifty guests in this civilized place is a vexation to you."

"I did not know they were scrutinizing," she answered; "and even if they are, what does it matter to me? I am sure I am quite too tired to care."

"Why have you come here?" asked the Disagreeable Man suddenly.

"Probably for the same reason as yourself," she said; "to get better or well."

"You won't get better," he answered cruelly; "I know your type well; you burn yourselves out quickly. And — my God — how I envy you!"

"So you have pronounced my doom," she said, looking at him intently. Then she laughed; but there was no merriment in the laughter.

"Listen," she said, as she bent nearer to him; "because you are hopeless, it does not follow that you should try to make others hopeless too. You have drunk deep of the cup of poison; I can see that. To hand the cup on to others is the part of a coward."

She walked past the English table, and the Polish table, and so out of the Kurhaus dining-hall.

CHAPTER II.

CONTAINS A FEW DETAILS.

In an old second-hand bookshop in London, an old man sat reading Gibbon's History of Rome. He did not put down his book when the postman brought him a letter. He just glanced indifferently at the letter, and impatiently at the postman. Zerviah Holme did not like to be interrupted when he was reading Gibbon; and as he was always reading Gibbon, an interruption was always regarded by him as an insult.

About two hours afterwards, he opened the letter, and learnt that his niece, Bernardine, had arrived safely in Petershof, and that she intended to get better and come home strong. He tore up the letter, and instinctively turned to the photograph on the mantelpiece. It was the picture of a face young and yet old, sad and yet with possibilities of merriment, thin and drawn and almost wrinkled, and with piercing eyes which, even in the dull lifelessness of the photograph, seemed to be burning themselves away. Not a pleasing nor a good face; yet intensely pathetic because of its undisguised harassment.

Zerviah looked at it for a moment.

"She has never been much to either of us," he said to himself. "And yet, when Malvina was alive, I used to think that she was hard on Bernardine. I believe I said so once or twice. But Malvina had her own way of looking at things. Well, that is over now."

He then, with characteristic speed, dismissed all thoughts which did not relate to Roman History; and the remembrance of Malvina, his wife, and Bernardine, his niece, took up an accustomed position in the background of his mind.

Bernardine had suffered a cheerless childhood in which dolls and toys took no leading part.

She had no affection to bestow on any doll, nor any woolly lamb, nor apparently on any human person; unless, perhaps, there was the possibility of a friendly inclination towards Uncle Zerviah, who would not have understood the value of any deeper feeling, and did not therefore call the child cold-hearted and unresponsive, as he might well have done.

This she certainly was, judged by the standard of other children; but then no softening influences had been at work during her tenderest years. Aunt Malvina knew as much about sympathy as she did about the properties of an ellipse; and even the fairies had failed to win little Bernardine. At first they tried with loving patience what they might do for her; they came out of their books, and danced and sang to her, and whispered sweet stories to her, at twilight, the fairies' own time. But she would have none of them, for all their gentle persuasion. So they gave up trying to please her, and left her as they had found her, loveless. What can be said of a childhood which even the fairies have failed to touch with the warm glow of affection?

Such a little restless spirit, striving to express itself now in this direction, now in that; yet always actuated by the same constant force, *the desire for work*. Bernardine seemed to have no special wish to be useful to others; she seemed just to have a natural tendency to work, even as others have a natural tendency to play. She was always in earnest; life for little Bernardine meant something serious.

Then the years went by. She grew up and filled her life with many interests and ambitions. She was at least a worker, if nothing else; she had always been a diligent scholar, and now she took her place as an able teacher. She was self-reliant, and, perhaps, somewhat conceited. But, at least, Bernardine the young woman had learnt something which Bernardine the young child had not been able to learn: she learnt how to smile. It took her about six and twenty years to learn; still, some people take longer than that; in fact, many never learn. This is a brief summary of Bernardine Holme's past.

Then, one day, when she was in the full swing

of her many engrossing occupations : teaching, writing articles for newspapers, attending socialistic meetings, and taking part in political discussions—she was essentially a "modern product," this Bernardine—one day she fell ill. She lingered in London for some time, and then she went to Petershof.

CHAPTER III.

MRS. REFFOLD LEARNS HER LESSON.

PETERSHOF was a winter resort for consumptive patients, though, indeed, many people who simply needed the change of a bracing climate went there to spend a few months; and came away wonderfully better for the mountain air. This was what Bernardine Holme hoped to do; she was broken down in every way, but it was thought that a prolonged stay in Petershof might help her back to a reasonable amount of health, or, at least, prevent her from slipping into further decline. She had come alone, because she had no relations except that old uncle, and no money to pay any friend who might have been willing to come with her. But she probably cared very little, and the morning after her arrival, she strolled out by herself, investigating the place where she was about to spend six months. She was dragging herself along, when she met the Disagreeable Man. She stopped him. He was not accustomed to be stopped by any one, and he looked rather astonished.

"You were not very cheering last night," she said to him.

"I believe I am not generally considered to be lively," he answered, as he knocked the snow off his boot.

"Still, I am sorry I spoke to you as I did," she went on frankly. "It was foolish of me to mind what you said."

He made no reference to his own remark, and was passing on his way again, when he turned back and walked with her.

"I have been here nearly seven years" he said, and there was a ring of sadness in his voice as he spoke, which he immediately corrected. "If you want to know anything about the place, I can tell you. If you are able to walk, I can show you some lovely spots, where you will not be bothered with people. I can take you to a snow fairy-land. If you are sad and disappointed, you will find shining comfort there. It is not all sadness in Petershof. In the silent snow forests, if you dig the snow away, you will find the tiny buds nestling in their white nursery. If the sun does not dazzle your eyes, you may always see the great mountains piercing the sky. These wonders have been a happiness to me. You are not too ill but that they may be a happiness to you also."

"Nothing can be much of a happiness to me," she said, half to herself, and her lips quivered. "I have had to give up so much: all my work, all my ambitions."

"You are not the only one who has had to do that," he said sharply. "Why make a fuss? Things arrange themselves, and eventually we adjust ourselves to the new arrangement. A great deal of caring and grieving, phase one; still more caring and grieving, phase two; less caring and grieving, phase three; no further feeling whatsoever, phase four. Mercifully I am at phase four. You are at phase one. Make a quick journey over the stages."

He turned and left her, and she strolled along, thinking of his words, wondering how long it would take her to arrive at his indifference. She had always looked upon indifference as paralysis of the soul, and paralysis meant death, nay, was worse than death. And here was this man, who had obviously suffered both mentally and physically, telling her that the only sensible course was to learn not to care. How could she learn not to care? All her life long she had studied and worked and cultivated herself in every direction in the hope of being able to take a high place in literature, or, in any case, to do something in life distinctly better than what other people did. When everything was coming near to her grasp, when there seemed a fair chance of realizing her ambitions, she had suddenly fallen ill, broken up so entirely in every way, that those who knew her when she was well, could scarcely recognize her now that she was ill. The doctors spoke of an overstrained nervous system: the pestilence of these modern days; they spoke of rest, change of work and scene, bracing air. She might regain her vitality; she might not. Those who had played themselves out must pay the penalty. She was thinking of her whole history, pitying herself profoundly, coming to the conclusion, after true human fashion, that she was the worst-used person on earth, and that no one but herself knew what disappointed ambitions were; she was thinking of all this, and looking profoundly miserable and martyr-like, when some one called her by her name. She looked round and saw one of the English ladies belonging to the Kurhaus; Bernardine had noticed her the previous night. She seemed in capital spirits, and had three or four admirers waiting on her very words. She was a tall, handsome woman, dressed in a superb fur-trimmed cloak, a woman of splendid bearing and address. Bernardine looked a contemptible little piece of humanity beside her. Some such impression conveyed itself to the two men who were walking with Mrs. Reffold. They looked at the one woman, and then at the other, and smiled at each other, as men do smile on such occasions.

"I am going to speak to this little thing," Mrs. Reffold had said to her two companions before they came near Bernardine. "I must find out who she is, and where she comes from. And, fancy, she has come quite alone. I have inquired. How hopelessly out of fashion she dresses. And what a hat!"

"I should not take the trouble to speak to her," said one of the men. "She may fasten herself on to you. You know what a bore that is."

"Oh, I can easily snub any one if I wish," replied Mrs. Reffold, rather disdainfully.

So she hastened up to Bernardine, and held out her well-gloved hand.

"I had not a chance of speaking to you last night, Miss Holme," she said. "You retired so early. I hope you have rested after your journey. You seemed quite worn out."

"Thank you," said Bernardine, looking admiringly at the beautiful woman, and envying her, just as all plain women envy their handsome sisters.

"You are not alone, I suppose?" continued Mrs. Reffold.

"Yes, quite alone," answered Bernardine.

"But you are evidently acquainted with **Mr. Allitsen**, your neighbour at table," said Mrs. Reffold; "so you will not feel quite lonely here. It is a great advantage to have a friend at a place like this."

"I never saw him before last night," said Bernardine.

"Is it possible?" said Mrs. Reffold, in her pleasantest voice. "Then you *have* made a triumph of the Disagreeable Man. He very rarely deigns to talk with any of us. He does not even appear to see us. He sits quietly and reads. It would be interesting to hear what his conversation is like. I should be quite amused to know what you did talk about."

"I dare say you would," said Bernardine quietly.

Then Mrs. Reffold, wishing to screen her inquisitiveness, plunged into a description of Petershof life, speaking enthusiastically about everything, except the scenery, which she did not mention. After a time she ventured to begin once more taking soundings. But somehow or other, those bright eyes of Bernardine, which looked at her so searchingly, made her a little nervous, and, perhaps, a little indiscreet.

"Your father will miss you," she said tentatively.

"I should think probably not," answered Bernardine. "One is not easily missed, you know." There was a twinkle in Bernardine's eye as she added, "He is probably occupied with other things."

"What is your father?" asked Mrs. Reffold, in her most coaxing tones.

"I don't know what he is now," answered Bernardine placidly. "But he was a genius. He is dead."

Mrs. Reffold gave a slight start, for she began to feel that this insignificant little person was making fun of her. This would never do, and before witnesses too. So she gathered together her best resources and said:

"Dear me, how very unfortunate: a genius too. Death is indeed cruel. And here one sees so much of it, that unless one learns to steel one's heart, one becomes melancholy. Ah, it is indeed sad to see all this suffering!" (Mrs. Reffold herself had quite succeeded in steeling her heart against her own invalid husband.) She then gave an account of several bad cases of consumption, not forgetting to mention two instances of suicide which had lately taken place in Petershof.

"One gentleman was a Russian," she said. "Fancy coming all the way from Russia to this little out-of-the-world place! But people come from the uttermost ends of the earth, though of course there are many Londoners here. I suppose you are from London?"

"I am not living in London now," said Bernardine cautiously.

"But you know it, without doubt," continued Mrs. Reffold. "There are several Kensington people here. You may meet some friends; indeed in our hotel there are two or three families from Lexham Gardens."

Bernardine smiled a little viciously; looked first at Mrs. Reffold's two companions with an amused sort of indulgence, and then at the lady herself. She paused a moment, and then said :

"Have you asked all the questions you wish to ask? And, if so, may I ask one of you? Where does one get the best tea?"

Mrs. Reffold gave an inward gasp, but pointed gracefully to a small confectionery shop on the other side of the road. Mrs. Reffold did everything gracefully.

Bernardine thanked her, crossed the road, and passed into the shop.

"Now I have taught her a lesson not to interfere with me," said Bernardine to herself. "How beautiful she is."

Mrs. Reffold and her two companions went silently on their way. At last the silence was broken.

"Well, I'm blessed!" said the taller of the two, lighting a cigar.

"So am I," said the other, lighting his cigar too.

"Those are precisely my own feelings," remarked Mrs. Reffold.

But she had learnt her lesson.

c 2

CHAPTER IV.

CONCERNING WÄRLI AND MARIE.

WÄRLI, the little hunchback postman, a cheery soul, came whistling up the Kurhaus stairs, carrying with him that precious parcel of registered letters, which gave him the position of being the most important person in Petershof. He was a linguist, too, was Wärli, and could speak broken English in a most fascinating way, agreeable to every one, but intelligible only to himself. Well, he came whistling up the stairs, when he heard Marie's blithe voice humming her favourite spinning-song.

"Ei, Ei!" he said to himself; "Marie is in a good temper to-day. I will give her a call as I pass."

He arranged his neckerchief and smoothed his curls; and when he reached the end of the landing, he paused outside a little glass-door, and, all unobserved, watched Marie in her pantry cleaning the candlesticks and lamps.

Marie heard a knock, and, looking up from her work, saw Wärli.

"Good day, Wärli," she said, glancing hurriedly at a tiny broken mirror suspended on the wall. "I suppose you have a letter for me. How delightful!"

"Never mind about the letter just now," he said, waving his hand as though wishing to dismiss the subject. "How nice to hear you singing so sweetly, Marie! Dear me, in the old days at Grüsch, how often I have heard that song of the spinning-wheels. You have forgotten the old days, Marie, though you remember the song."

"Give me my letter, Wärli, and go about your work," said Marie, pretending to be impatient. But all the same her eyes looked extremely friendly. There was something very winning about the hunchback's face.

"Ah, ah! Marie," he said, shaking his curly head; "I know how it is with you: you only like people in fine binding. They have not always fine hearts."

"What nonsense you talk, Wärli!" said Marie. "There, just hand me the oil-can. You can fill this lamp for me. Not too full, you goose! And this one also; ah, you're letting the oil trickle down! Why, you're not fit for anything except carrying letters! Here, give me my letter."

"What pretty flowers," said Wärli. "Now if there is one thing I do like, it is a flower. Can you spare me one, Marie? Put one in my button-hole, do!"

"You are a nuisance this afternoon," said Marie, smiling and pinning a flower on Wärli's blue coat. Just then a bell rang violently.

"Those Portuguese ladies will drive me quite mad," said Marie. "They always ring just when I am enjoying myself."

"When you are enjoying yourself!" said Wärli triumphantly.

"Of course," returned Marie; "I always do enjoy cleaning the oil-lamps; I always did!"

"Ah, I d forgotten the oil-lamps!" said Wärli.

"And so had I!" laughed Marie. "Na, na, there goes that bell again! Won't they be angry! Won't they scold at me! Here, Wärli, give me my letter, and I'll be off."

"I never told you I had any letter for you," remarked Wärli. "It was entirely your own idea. Good afternoon, Fräulein Marie."

The Portuguese ladies' bell rang again, still more passionately this time; but Marie did not seem to hear nor care. She wished to be revenged on that impudent postman. She went to the top of the stairs and called after Wärli in her most coaxing tones:

"Do step down one moment; I want to show you something!"

"I must deliver the registered letters," said Wärli, with official haughtiness. "I have already wasted too much of my time."

"Won't you waste a few more minutes on me?" pleaded Marie pathetically. "It is not often I see you now."

Wärli came down again, looking very happy.

"I want to show you such a beautiful photograph I've had taken," said Marie. "Ach, it is beautiful!"

"You must give one to me," said Wärli eagerly.

"Oh, I can't do that," replied Marie, as she opened the drawer and took out a small packet. "It was a present to me from the Polish gentleman himself. He saw me the other day here in the pantry. I was so tired, and I had fallen asleep, with my broom, just as you see me here. So he made a photograph of me. He admires me very much. Isn't it nice? and isn't the Polish gentleman clever? and isn't it nice to have so much attention paid to one? Oh, there's that horrid bell again! Good afternoon, Herr Wärli. That is all I have to say to you, thank you."

Wärli's feelings towards the Polish gentleman were not of the friendliest that day.

CHAPTER V.

THE DISAGREEABLE MAN.

ROBERT ALLITSEN told Bernardine that she was not likely to be on friendly terms with the English people in the Kurhaus.

"They will not care about you, and you will not care about the foreigners. So you will thus be thrown on your own resources, just as I was when I came."

"I cannot say that I have any resources," Bernardine answered. "I don't feel well enough to try to do any writing, or else it would be delightful to have the uninterrupted leisure."

So she had probably told him a little about her life and occupation; although it was not likely that she would have given him any serious confidences. Still, people are often surprisingly frank about themselves, even those who pride themselves upon being the most reticent mortals in the world.

"But now, having the leisure," she continued, "I have not the brains."

"I never knew any writer who had," said the Disagreeable Man grimly.

"Perhaps your experience has been limited," she suggested.

"Why don't you read?" he said. "There is a good library here. It contains all the books we don't want to read."

"I am tired of reading," Bernardine said. "I seem to have been reading all my life. My uncle, with whom I live, keeps a second-hand book-shop, and ever since I can remember, I have been surrounded by books. They have not done me much good, nor any one else either."

"No, probably not," he said. "But now that you have left off reading, you will have a chance of learning something, if you live long enough. It is wonderful how much one does learn when one does not read. It is almost awful. If you don't care about reading now, why do you not occupy yourself with cheese-mites?"

"I do not feel drawn towards cheese-mites."

"Perhaps not, at first; but all the same they form a subject which is very engaging. Or any branch of bacteriology."

"Well, if you were to lend me your microscope, perhaps I might begin."

"I could not do that," he answered quickly. "I never lend my things."

"No, I did not suppose you would," she said. "I knew I was safe in making the suggestion."

"You are rather quick of perception in spite of all your book reading," he said. "Yes, you are quite right. I am selfish. I dislike lending my things, and I dislike spending my money except on myself. If you have the misfortune to linger on as I do, you will know that it is perfectly legitimate to be selfish in small things, *if one has made the one great sacrifice.*"

"And what may that be?"

She asked so eagerly that he looked at her,

and then saw how worn and tired her face was; and the words which he was intending to speak, died on his lips.

"Look at those asses of people on toboggans," he said brusquely. "Could you manage to enjoy yourself in that way? That might do you good."

"Yes," she said; "but it would not be any pleasure to me."

She stopped to watch the toboggans flying down the road. And the Disagreeable Man went his own solitary way, a forlorn figure, with a face almost expressionless, and a manner wholly impenetrable.

He had lived nearly seven years at Petershof, and, like many others, was obliged to continue staying there if he wished to continue staying in this planet. It was not probable that he had any wish to prolong his frail existence, but he did his duty to his mother by conserving his life; and this feeble flame of duty and affection was the only lingering bit of warmth in a heart frozen almost by ill health and disappointed ambitions. The moralists tell us that suffering ennobles, and that a right acceptation of hindrances goes towards forming a beautiful character. But this result must largely depend on the original character: certainly, in the case of Robert Allitsen, suffering had not ennobled his mind, nor disappointment sweetened his disposition. His title of "Disagreeable Man" had been fairly earned, and he hugged it to himself with a triumphant secret satisfaction.

There were some people in Petershof who were inclined to believe certain absurd rumours about his alleged kindness. It was said that on more than one occasion he had nursed the suffering and the dying in sad Petershof, and, with all the sorrowful tenderness worthy of a loving mother, had helped them to take their leave of life. But these were only rumours, and there was nothing in Robert Allitsen's ordinary bearing to justify such talk. So the foolish people who, for the sake of making themselves peculiar, revived these unlikely fictions, were speedily ridiculed and reduced to silence. And the Disagreeable Man remained the Disagreeable Man, with a clean record for unamiability.

He lived a life apart from others. Most of his time was occupied in photography, or in the use and study of the microscope, or in chemistry. His photographs were considered to be most beautiful. Not that he showed them specially to any one; but he generally sent a specimen of his work to the Monthly Photograph Portfolio, and hence it was that people learned to know of his skill. He might be seen any fine day trudging along in company with his photographic apparatus, and a desolate dog, who looked almost as cheerless as his chosen comrade. Neither the one took any notice of the other; Allitsen was no more genial to the dog than he was to the Kurhaus guests; the dog was no more demonstrative to Robert Allitsen than he was to any one in Petershof.

Still, they were "something" to each other: that unexplainable "something" which has to explain almost every kind of attachment.

He had no friends in Petershof, and apparently had no friends anywhere. No one wrote to him, except his old mother; the papers which were sent to him came from a stationer's.

He read all during meal-time. But now and again he spoke a few words with Bernardine Holme, whose place was next to him. It never occurred to him to say good morning, nor to give a greeting of any kind, nor to show a courtesy. One day during lunch, however, he did take the trouble to stoop and pick up Bernardine Holme's shawl, which had fallen for the third time to the ground.

"I never saw a female wear a shawl more carelessly than you," he said. "You don't seem to know anything about it."

His manner was always gruff. Every one complained of him. Every one always had complained of him. He had never been heard to laugh. Once or twice he had been seen to smile on occasions when people talked confidently of recovering their health. It was a beautiful smile worthy of a better cause. It was a smile which made one pause to wonder what could have been the original disposition of the Disagreeable Man before ill-health had cut him off from the affairs of active life. Was he happy or unhappy? It was not known. He gave no sign of either the one state or the other. He always looked very ill, but he did not seem to get worse. He had never been known to make the faintest allusion to his own health. He never "smoked" his thermometer in public; and this was the more remarkable in an hotel where people would even leave off a conversation and say: "Excuse me, Sir or Madam, I must now take my temperature. We will resume the topic in a few minutes."

He never lent any papers or books; and he never borrowed any.

He had a room at the top of the hotel, and he lived his life, amongst his chemistry bottles, his scientific books, his microscope, and his camera. He never sat in any of the hotel drawing-rooms. There was nothing striking nor eccentric about his appearance. He was neither ugly nor good-looking, neither tall nor short, neither fair nor dark. He was thin and frail, and rather bent. But that might be the description of any one in Petershof. There was nothing pathetic about him, no suggestion even of poetry, which gives a reverence to suffering, whether mental or physical. As there was no expression on his face, so also there was no expression in his eyes: no distant longing, no far-off fixedness; nothing, indeed, to awaken sad sympathy.

The only positive thing about him was his rudeness. Was it natural or cultivated? No one in Petershof could say. He had always been as he was; and there was no reason to suppose that he would ever be different.

He was, in fact, like the glacier of which he had such a fine view from his room; like the glacier, an unchanging feature of the neighbourhood.

No one loved it better than the Disagreeable Man did; he watched the sunlight on it, now pale golden, now fiery red. He loved the sky, the dull grey, or the bright blue. He loved the snow forests, and the snow-girt streams, and the ice cathedrals, and the great firs patient beneath their snow-burden. He loved the frozen waterfalls, and the costly

D

diamonds in the snow. He knew, too, where the flowers nestled in their white nursery. He was, indeed, an authority on Alpine botany. The same tender hands which plucked the flowers in the spring-time, dissected them and laid them bare beneath the microscope. But he did not love them the less for that.

Were these pursuits a comfort to him? Did they help him to forget that there was a time when he, too, was burning with ambition to distinguish himself, and be one of the marked men of the age?

Who could say?

CHAPTER VI.

THE TRAVELLER AND THE TEMPLE OF KNOWLEDGE.

Countless ages ago a Traveller, much worn with journeying, climbed up the last bit of rough road which led to the summit of a high mountain. There was a temple on that mountain. And the Traveller had vowed that he would reach it before death prevented him. He knew the journey was long, and the road rough. He knew that the mountain was the most difficult of ascent of that mountain chain, called "The Ideals." But he had a strongly-hoping heart and a sure foot. He lost all sense of time, but he never lost the feeling of hope.

"Even if I faint by the way-side," he said to

himself, "and am not able to reach the summit, still it is something to be on the road which leads to the High Ideals."

That was how he comforted himself when he was weary. He never lost more hope than that; and surely that was little enough.

And now he had reached the temple.

He rang the bell, and an old white-haired man opened the gate. He smiled sadly when he saw the Traveller.

"*And yet another one,*" he murmured. "What does it all mean?"

The Traveller did not hear what he murmured.

"Old white-haired man," he said, "tell me; and so I have come at last to the wonderful Temple of Knowledge. I have been journeying hither all my life. Ah, but it is hard work climbing up to the Ideals."

The old man touched the Traveller on the arm. "Listen," he said gently. "This is not the Temple of Knowledge. And the Ideals are not a chain of mountains; they are a stretch of plains, and the Temple of Knowledge is in their centre. You have come the wrong road. Alas, poor Traveller!"

The light in the Traveller's eyes had faded. The hope in his heart died. And he became old and withered. He leaned heavily on his staff.

"Can one rest here?" he asked wearily.

"No."

"Is there a way down the other side of these mountains?"

"No."

"What are these mountains called?"

"They have no name."

"And the temple — how do you call the temple?"

"It has no name."

"Then I call it the Temple of Broken Hearts," said the Traveller."

And he turned and went. But the old white-haired man followed him.

"Brother," he said, "you are not the first to come here, but you may be the last. Go back to the plains, and tell the dwellers in the plains that the Temple of True Knowledge is in their very midst; any one may enter it who chooses; the gate is not even closed. The Temple has always been in the plains, in the very heart of life, and work, and daily effort. The philosopher may enter, the stone-breaker may enter. You must have passed it every day of your life; a plain, venerable building, unlike your glorious cathedrals."

"I have seen the children playing near it," said the Traveller. "When I was a child I used to play there. Ah, if I had only known! Well, the past is the past."

He would have rested against a huge stone, but that the old white-haired man prevented him.

"Do not rest," he said. "If you once rest there, you will not rise again. When you once rest, you will know how weary you are."

"I have no wish to go farther," said the Traveller. "My journey is done; it may have been in the wrong direction, but still it is done."

"Nay, do not linger here," urged the old

man. "Retrace your steps. Though you are broken-hearted yourself, you may save others from breaking their hearts. Those whom you meet on this road, you can turn back. Those who are but starting in this direction you can bid pause and consider how mad it is to suppose that the Temple of True Knowledge should have been built on an isolated and dangerous mountain. Tell them that although God seems hard, He is not as hard as all that. Tell them that the Ideals are not a mountain range, but their own plains, where their great cities are built, and where the corn grows, and where men and women are toiling, sometimes in sorrow and sometimes in joy."

"I will go," said the Traveller.

And he started.

But he had grown old and weary. And the journey was long; and the retracing of one's steps is more toilsome than the tracing of them. The ascent, with all the vigour and hope of life to help him, had been difficult enough; the descent, with no vigour and no hope to help him, was almost impossible.

So that it was not probable that the Traveller lived to reach the plains. But whether he reached them or not, still he had started

And not many Travellers do that.

CHAPTER VII.

BERNARDINE.

The crisp mountain air and the warm sunshine began slowly to have their effect on Bernardine, in spite of the Disagreeable Man's verdict. She still looked singularly lifeless, and appeared to drag herself about with painful effort; but the place suited her, and she enjoyed sitting in the sun listening to the music which was played by a scratchy string band. Some of the Kurhaus guests, seeing that she was alone and ailing, made some attempts to be kindly to her. She always seemed astonished that people should concern themselves about her; whatever her faults were, it never struck her that she might be of any importance to others, however important she might be to herself. She was grateful for any little kindness which was shewn her; but at first she kept very much to herself, talking chiefly with the Disagreeable Man, who, by the way, had surprised every one —but no one more than himself—by his unwonted behaviour in bestowing even a fraction of his companionship on a Petershof human being.

There was a great deal of curiosity about her, but no one ventured to question her since Mrs. Reffold's defeat. Mrs. Reffold herself rather avoided her, having always a vague suspicion that Bernardine tried to make fun of her. But whether out of perversity or not, Bernardine never would be avoided by her, never let her pass by without a few words of conversation, and always went to her for information, much to the amusement of Mrs. Reffold's faithful attendants. There was always a twinkle in Bernardine's eye when she spoke with Mrs. Reffold. She never fastened herself on to any one; no one could say she intruded. As time went, on there was a vague sort of feeling that she did not intrude enough. She was ready to speak if any one cared to speak with her, but she never began a conversation except with Mrs. Reffold. When people did talk to her, they found her genial. Then the sad face would smile kindly, and the sad eyes speak kind sympathy. Or some bit of fun would flash forth, and a peal of young laughter ring out. It seemed strange that such fun could come from her.

Those who noticed her, said she appeared always to be thinking.

She was thinking and learning.

Some few remarks roughly made by the Disagreeable Man had impressed her deeply.

"You have come to a new world," he said. "the world of suffering. You are in a fury because your career has been checked, and because you have been put on the shelf; you, of all people. Now you will learn how many quite as able as yourself, and abler, have been put on the shelf too, and have to stay there. You are only a pupil in suffering. What about the professors? If your wonderful wisdom has left you with any sense at all, look about you and learn."

So she was looking, and thinking, and learning. And as the days went by, perhaps a softer light came into her eyes.

All her life long, her standard of judging people had been an intellectual standard, or an artistic standard: what people had done with outward and visible signs; how far they had contributed to thought; how far they had influenced any great movement, or originated it; how much of a benefit they had been to their century or their country; how much social or political activity, how much educational energy they had devoted to the pressing need of the times.

She was undoubtedly a clever, cultured young woman; the great work of her life had been self-culture. To know and understand, she had spared neither herself nor any one else. To know, and to use her acquired knowledge intellectually as teacher and, perhaps, too, as writer, had been the great aim of her life. Everything that furthered this aim won her instant attention. It never struck her that she was selfish. One does not think of that until the great check comes. One goes on, and would go on. But a barrier rises up. Then, finding one can advance no further, one turns round; and what does one see?

Bernardine saw that she had come a long journey. She saw what the Traveller saw. That was all she saw at first. Then she remembered that she had done the journey entirely for her own sake. Perhaps it might not have looked so dreary if it had been undertaken for some one else.

She had claimed nothing of any one; she had given nothing to any one. She had simply taken her life in her own hands and made what she could of it. What had she made of it?

Many women asked for riches, for position, for influence and authority and admiration. She had only asked to be able to work. It seemed little enough to ask. That she asked so little placed her, so she thought, apart from the common herd of eager askers. To be cut off from active life and earnest work was a possibility which never occurred to her.

It never crossed her mind that in asking for the one thing for which she longed, she was really asking for the greatest thing. Now, in the hour of her enfeeblement, and in the hour of the bitterness of her heart, she still prided herself upon wanting so little.

"It seems so little to ask," she cried to herself time after time. "I only want to be able to do a few strokes of work. I would be content now to do so little, if only I might do some. The laziest day-labourer on the road would laugh at the small amount of work which would content me now."

She told the Disagreeable Man that one day.

"So you think you are moderate in your demands," he said to her. "You are a most amusing young woman. You are so perfectly unconscious how exacting you really are. For, after all, what is it you want? You want to have that wonderful brain of yours restored, so that you may begin to teach, and, perhaps, write a book. Well, to repeat my former words: you are still at phase one, and you are longing to be strong enough to fulfil your ambitions and write a book. When you arrive at phase four, you will be quite content to dust one of your uncle's books instead: far more useful work and far more worthy of encouragement. If every one who wrote books now would be satisfied to dust books already written, what a regenerated world it would become!"

She laughed good-temperedly. His remarks did not vex her; or, at least, she showed no vexation. He seemed to have constituted himself as her critic, and she made no objections. She had given him little bits of stray confidence about herself, and she received everything he had to say with that kind of forbearance which chivalry bids us show to the weak and ailing. She made allowances for him; but she did more than that for him: she did not let him see that she made allowances. Moreover, she recognized amidst all his roughness a certain kind of sympathy which she could not resent, because it was not aggressive. For to some natures the expression of sympathy is an irritation; to be sympathized with means to be pitied, and to be pitied means to be looked down upon. She was sorry for him, but she would not have told him so for worlds; he would have shrunk from pity as much as she did. And yet the sympathy which she thought she did not want for herself, she was silently giving to those around her, like herself, thwarted, each in a different way perhaps, still thwarted all the same.

She found more than once that she was learning to measure people by a standard different from her former one; not by what they had *done* or *been*, but by what they had *suffered*. But such a change as this does not come suddenly, though, in a place like Petershof, it comes quickly, almost unconsciously.

She became immensely interested in some of the guests; and there were curious types in the Kurhaus. The foreigners attracted her chiefly; a little Parisian danseuse, none too quiet in her manner, won Bernardine's fancy.

"I so want to get better, *chérie*," she said to Bernardine. "Life is so bright. Death: ah, how the very thought makes one shiver! That horrid doctor says I must not skate; it is not wise. When was I wise? Wise people don't enjoy themselves. And I have enjoyed myself, and will still."

"How can you go about with that little danseuse?" the Disagreeable Man said to Bernardine one day. "Do you know who she is?"

"Yes," said Bernardine; "she is the lady who thinks you must be a very ill-bred person because you stalk into meals, with your hands in your pockets. She wondered how I could bring myself to speak to you."

"I dare say many people wonder at that," said Robert Allitsen rather peevishly.

"Oh no," replied Bernardine; "they wonder that you talk to me. They think I must either be very clever or else very disagreeable."

"I should not call you clever," said Robert Allitsen grimly.

"No," answered Bernardine pensively. "But I always did think myself clever until I came here. Now I am beginning to know better. But it is rather a shock, isn't it?"

"I have never experienced the shock," he said.

E

"Then you still think you are clever?" she asked.

"There is only one man my intellectual equal in Petershof, and he is not here any more," he said gravely. "Now I come to remember, he died. That is the worst of making friendships here; people die."

"Still, it is something to be left king of the intellectual world," said Bernardine. "I never thought of you in that light."

There was a sly smile about her lips as she spoke, and there was the ghost of a smile on the Disagreeable Man's face.

"Why do you talk with that horrid Swede?" he said suddenly. "He is a wretched low foreigner. Have you heard some of his views?"

"Some of them," answered Bernardine cheerfully. "One of his views is really amusing: that it is very rude of you to read the newspaper during meal-time; and he asks if it is an English custom. I tell him it depends entirely on the Englishman, and the Englishman's neighbour!"

So she too had her raps at him, but always in the kindest way.

He had a curious effect on her. His very bitterness seemed to check in its growth her own bitterness. The cup of poison of which he himself had drunk deep, he passed on to her. She drank of it, and it did not poison her. She was morbid, and she needed cheerful companionship. His dismal companionship and his hard way of looking at life ought by rights to have oppressed her. Instead of which she became less sorrowful.

Was the Disagreeable Man, perhaps, a reader of character? Did he know how to help her in his own grim gruff way? He himself had suffered so much; perhaps he did know.

CHAPTER VIII.

THE STORY MOVES ON AT LAST.

Bernardine was playing chess one day with the Swedish Professor. On the Kurhaus terrace the guests were sunning themselves, warmly wrapped up to protect themselves from the cold, and well-provided with parasols to protect themselves from the glare. Some were reading, some were playing cards or Russian dominoes, and others were doing nothing. There was a good deal of fun, and a great deal of screaming amongst the Portuguese colony. The little danseuse and three gentlemen acquaintances were drinking coffee, and not behaving too quietly. Pretty Fräulein Müller was leaning over her balcony carrying on a conversation with a picturesque Spanish youth below. Most of the English party had gone sledging and tobogganing. Mrs. Reffold had asked Bernardine to join them, but she had refused. Mrs. Reffold's friends were anything but attractive to Bernardine, although she liked Mrs. Reffold herself immensely. There was no special reason why she should like her; she certainly had no cause to admire her every-day behaviour, nor her neglect of her invalid husband, who was passing away, uncared for in the present, and not likely to be mourned for in the future. Mrs. Reffold was gay, careless, and beautiful. She understood nothing about nursing, and cared less. So a trained nurse looked after Mr. Reffold, and Mrs. Reffold went sledging.

"Dear Wilfrid is so unselfish," she said. "He will not have me stay at home. But I feel very selfish." That was her stock remark. Most people answered her by saying: "Oh no, Mrs. Reffold, don't say that." But when she made the remark to Bernardine, and expected the usual reply, Bernardine said instead:

"Mr. Reffold seems lonely."

"Oh, he has a trained nurse, and she can read to him," said Mrs. Reffold hurriedly. She seemed ruffled.

"I had a trained nurse once," replied Bernardine; "and she could read; but she would not. She said it hurt her throat."

"Dear me, how very unfortunate for you," said Mrs. Reffold. "Ah, there is Captain Graham calling. I must not keep the sledges waiting."

That was a few days ago, but to-day, when Bernardine was playing chess with the Swedish Professor, Mrs. Reffold came to her. There was a curious mixture of shyness and abandon in Mrs. Reffold's manner.

"Miss Holme," she said, "I have thought of such a splendid idea. Will you go and see Mr. Reffold this afternoon? That would be a nice little change for him."

Bernardine smiled.

"If you wish it," she answered.

Mrs. Reffold nodded and hastened away, and

Bernardine continued her game, and, having finished it, rose to go.

The Reffolds were rich, and lived in a suite of apartments in the more luxurious part of the Kurhaus. Bernardine knocked at the door, and the nurse came to open it.

"Mrs. Reffold asks me to visit Mr. Reffold," Bernardine said; and the nurse showed her into the pleasant sitting-room.

Mr. Reffold was lying on the sofa. He looked up as Bernardine came in, and a smile of pleasure spread over his wan face.

"I don't know whether I intrude," said Bernardine; "but Mrs. Reffold said I might come to see you."

Mr. Reffold signed to the nurse to withdraw.

She had never before spoken to him. She had often seen him lying by himself in the sunshine.

"Are you paid for coming to me?" he asked eagerly.

The words seemed rude enough, but there was no rudeness in the manner.

"No, I am not paid," she said gently; and then she took a chair and sat near him.

"Ah, that's well!" he said, with a sigh of relief. "I'm so tired of paid service. To know that things are done for me because a certain amount of francs are given so that those things may be done—well, one gets weary of it; that's all!"

There was bitterness in every word he spoke. "I lie here," he said, "and the loneliness of it—the loneliness of it!"

"Shall I read to you?" she asked kindly. She did not know what to say to him.

"I want to talk first," he replied. "I want to talk first to some one who is not paid for talking to me. I have often watched you, and wondered who you were. Why do you look so sad? No one is waiting for you to die?"

"Don't talk like that!" she said; and she bent over him and arranged the cushions for him more comfortably. He looked just like a great lank tired child.

"Are you one of my wife's friends?" he asked

"I don't suppose I am," she answered gently; "but I like her, all the same. Indeed, I like her very much. And I think her beautiful."

"Ah, she is beautiful!" he said eagerly. "Doesn't she look splendid in her furs? By Jove, you are right! She is a beautiful woman. I am proud of her."

Then the smile faded from his face.

"Beautiful," he said half to himself, "but hard."

"Come now," said Bernardine; "you are surrounded with books and newspapers. What shall I read to you?"

"No one reads what I want," he answered peevishly. "My tastes are not their tastes. I don't suppose you would care to read what I want to hear."

"Well," she said cheerily, "try me. Make your choice."

"Very well, the *Sporting and Dramatic*," he said. "Read every word of that. And about that theatrical divorce case. And every word of that too. Don't you skip, and cheat me."

She laughed and settled herself down to amuse him. And he listened contentedly.

"That is something like literature," he said once or twice. "I can understand papers of that sort going like wild-fire."

When he was tired of being read to, she talked to him in a manner that would have astonished the Disagreeable Man: not of books, nor learning, but of people she had met and of places she had seen; and there was fun in everything she said. She knew London well, and she could tell him about the Jewish and the Chinese quarters, and about her adventures in company with a man who took her here, there, and everywhere.

She made him some tea, and she cheered the poor fellow as he had not been cheered for months.

"You're just a little brick!" he said, when she was leaving. Then once more he added eagerly:

"And you're not to be paid, are you?"

"Not a single *sou!*" she laughed. "What a strange idea of yours!"

"You are not offended?" he said anxiously "But you can't think what a difference it makes to me. You are not offended?"

"Not in the least!" she answered. "I know quite well how you mean it. You want a little kindness with nothing at the back of it. Now, good-bye!"

He called her when she was outside the door.

"I say, will you come again soon?"

"Yes, I will come to-morrow."

"Do you know you've been a little brick. I hope I haven't tired you. You are only a bit of a thing yourself. But, by Jove, you know how to put a fellow in a good temper!"

When Mrs. Reffold went down to *table-d'hôte* that night, she met Bernardine on the stairs, and stopped to speak with her.

"We've had a splendid afternoon," she said; "and we've arranged to go again to-morrow at the same time. Such a pity you don't come! Oh, by the way, thank you for going to see my husband. I hope he did not tire you. He is a little querulous, I think. He so enjoyed your visit. Poor fellow! it is sad to see him so ill, isn't it?"

CHAPTER IX.

BERNARDINE PREACHES.

AFTER this, scarcely a day passed but Bernardine went to see Mr. Reffold. The most inexperienced eye could have known that he was becoming rapidly worse. Marie, the chambermaid, knew it, and spoke of it frequently to Bernardine.

"The poor lonely fellow!" she said, time after time.

Every one, except Mrs. Reffold, seemed to recognize that Mr. Reffold's days were numbered. Either she did not or would not understand. She made no alteration in the disposal of her time: sledging parties and skating picnics were the order of the day; she was thoroughly pleased with herself, and received the attentions of her admirers as a matter of course. The Petershof climate had got into her head; and it is a well-known fact that this glorious air has the effect on some people of banishing from their minds all inconvenient notions of duty and devotion, and all memory of the special object of their sojourn in Petershof. The coolness and calmness with which such people ignore their responsibilities, or allow strangers to assume them, would be an occasion for humour, if it were not an opportunity for indignation: though indeed it would take a very exceptionally sober-minded spectator not to get some fun out of the blissful self-satisfaction and unconsciousness which characterize the most negligent of 'caretakers.'

Mrs. Reffold was not the only sinner in this respect. It would have been interesting to get together a tea-party of invalids alone, and set the ball rolling about the respective behaviours of their respective friends. Not a pleasing chronicle: no very choice pages to add to the book of real life; still, valuable items in their way, representative of the actual as opposed to the ideal. In most instances there would have been ample testimony to that cruel monster known as Neglect.

Bernardine spoke once to the Disagreeable Man on this subject. She spoke with indignation, and he answered with indifference, shrugging his shoulders.

"These things occur," he said. "It is not that they are worse here than everywhere else; it is simply that they are together in an accumulated mass, and, as such, strike us with tremendous force. I myself am accustomed to these exhibitions of selfishness and neglect. I should be astonished if they did not take place. Don't mix yourself up with anything. If people are neglected, they *are* neglected, and there is the end of it. To imagine that you or I are going to do any good by filling up the breach, is simply an insanity leading to unnecessarily disagreeable consequences. I know you go to see Mr. Reffold. Take my advice, and keep away.

"You speak like a Calvinist," she answered, rather ruffled, "with the quintessence of self-protectiveness; and I don't believe you mean a word you say."

"My dear young woman," he said, "we are not living in a poetry book bound with gilt edges. We are living in a paper-backed volume of prose. Be sensible. Don't ruffle yourself on account of other people. Don't even trouble to criticize them; it is only a nuisance to yourself. All this simply points back to my first suggestion: fill up your time with some hobby, cheese-mites or the influenza bacillus, and then you will be quite content to let people be neglected, lonely, and to die. You will look upon it as an ordinary and natural process."

She waved her hand as though to stop him.

"There are days," she said, "when I can't bear to talk with you. And this is one of them."

"I am sorry," he answered, quite gently for him. And he moved away from her, and started for his usual lonely walk.

Bernardine turned home, intending to go to

see Mr. Reffold. He had become quite attached to her, and looked forward eagerly to her visits. He said her voice was gentle and her manner quiet; there was no bustling vitality about her to irritate his worn nerves. He was probably an empty-headed, stupid fellow; but it was none the less sad to see him passing away.

He called her 'Little Brick.' He said that no other epithet suited her so exactly. He was quite satisfied now that she was not paid for coming to see him. As for the reading, no one could read the *Sporting and Dramatic News* and the *Era* so well as Little Brick. Sometimes he spoke with her about his wife, but only in general terms of bitterness, and not always complainingly. She listened and said nothing.

"I'm a chap that wants very little," he said once. "Those who want little, get nothing."

That was all he said, but Bernardine knew to whom he referred.

To-day, as Bernardine was on her way back to the Kurhaus, she was thinking constantly of Mrs. Reffold, and wondering whether she ought to be made to realize that her husband was becoming rapidly worse. Whilst engrossed with this thought, a long train of sledges and toboggans passed her. The sound of the bells and the noisy merriment made her look up, and she saw beautiful Mrs. Reffold amongst the pleasure-seekers.

"If only I dared tell her now," said Bernardine to herself, "loudly and before them all."

Then a more sensible mood came over her.

"After all, it is not my affair," she said.

And the sledges passed away out of hearing.

When Bernardine sat with Mr. Reffold that afternoon she did not mention that she had seen his wife. He coughed a great deal, and seemed to be worse than usual, and complained of fever. But he liked to have her, and would not hear of her going.

"Stay," he said. "It is not much of a pleasure to you, but it is a great pleasure to me."

There was an anxious look on his face,

such a look as people wear when they wish to ask some question of great moment, but dare not begin.

At last he seemed to summon up courage.

"Little Brick," he said, in a weak, low voice, "I have something on my mind. You won't laugh, I know. You're not the sort. I know you're clever and thoughtful, and all that; you could tell me more than all the parsons put together. I know you're clever; my wife says so. She says only a very clever woman would wear such boots and hats."

Bernardine smiled.

"Well," she said kindly, "tell me."

"You must have thought a good deal, I suppose," he continued, "about life and death, and that sort of thing. I've never thought at all. Does it matter, Little Brick? It's too late now, I can't begin to think. But speak to me; tell me what you think. Do you believe we get another chance, and are glad to behave less like curs and brutes? Or is it all ended in that lonely little churchyard here? I've never troubled about these things before, but now

I know I am so near that gloomy little churchyard—well, it makes me wonder. As for the Bible, I never cared to read it. I was never much of a reader, though I've got through two or three firework novels and sporting stories. Does it matter, Little Brick?"

"How do I know?" she said gently. "How does any one know? People say they know; but it is all a great mystery—nothing but a mystery. Everything that we say, can be but a guess. People have gone mad over their guessing, or they have broken their hearts. But still the mystery remains, and we cannot solve it."

"If you don't know anything, Little Brick," he said, "at least tell me what you think: and don't be too learned; remember I'm only a brainless fellow."

He seemed to be waiting eagerly for her answer.

"If I were you," she said, "I should not worry. Just make up your mind to do better when you get another chance. One can't do more than that. That is what I shall think of: that God will give each one of us another chance, and that each one of us will take it and do better—I and you and every one. So there is no need to fret over failure, when one hopes one may be allowed to redeem that failure later on. Besides which, life is very hard. Why, we ourselves recognize that. If there be a God, some Intelligence greater than human intelligence, he will understand better than ourselves that life is very hard and difficult, and he will be astonished not *because we are not better, but because we are not worse.* At least, that would be my notion of a God. I should not worry, if I were you. Just make up your mind to do better if you get the chance, and be content with that."

"If that is what you think, Little Brick," he answered, "it is quite good enough for me And it does not matter about prayers and the Bible, and all that sort of thing?"

"*I* don't think it matters," she said. "I never have thought such things mattered. What does matter, is to judge gently, and not to come down like a sledge-hammer on other people's failings. Who are we, any of us, that we should be hard on others?"

"And not come down like a sledge-hammer on other people's failings," he repeated slowly. "I wonder if I have ever judged gently."

"I believe you have," she answered.

He shook his head.

"No," he said; "I have been a paltry fellow. I have been lying here, and elsewhere too, eating my heart away with bitterness, until you came. Since then I have sometimes forgotten to feel bitter. A little kindness does away with a great deal of bitterness."

He turned wearily on his side.

"I think I could sleep, Little Brick," he said, almost in a whisper. "I want to dream about your sermon. And I'm not to worry, am I?"

"No," she answered, as she stepped noiselessly across the room; "you are not to worry."

CHAPTER X.

THE DISAGREEABLE MAN IS SEEN IN A NEW LIGHT.

One specially fine morning a knock came at Bernardine's door. She opened it, and found Robert Allitsen standing there, trying to recover his breath.

"I am going to Loschwitz, a village about twelve miles off," he said. "And I have ordered a sledge. Do you care to come too?"

"If I may pay my share," she said.

"Of course," he answered; "I did not suppose you would like to be paid for any better than I should like to pay for you."

Bernardine laughed.

"When do we start?" she asked.

"Now," he answered. "Bring a rug, and also that shawl of yours which is always falling down, and come at once without any fuss. We shall be out for the whole day. What about Mrs. Grundy? We could manage to take her if you wished, but she would not be comfortable sitting amongst the photographic apparatus, and I certainly should not give up my seat to her."

"Then leave her at home," said Bernardine cheerily.

And so they settled it.

In less than a quarter of an hour they had started; and Bernardine leaned luxuriously back to enjoy to the full her first sledge-drive.

It was all new to her: the swift passing through the crisp air without any sensation of motion; the sleepy tinkling of the bells on the horses' heads; the noiseless cutting through of the snow-path.

All these weeks she had known nothing of the country, and now she found herself in the snow fairy-land of which the Disagreeable Man had often spoken to her. Around, vast plains of untouched snow, whiter than any dream of whiteness, jewelled by the sunshine with priceless diamonds, numberless as the sands of the sea. The great pines bearing their burden of snow patiently; others, less patient, having shaken themselves free from what the heavens had sent them to bear. And now the streams, flowing on reluctantly over ice-coated rocks, and the ice cathedrals formed by the icicles between the rocks.

And always the same silence, save for the tinkling of the horses' bells.

On the heights the quaint châlets, some merely huts for storing wood; on others, farms, or the homes of peasants; some dark brown, almost black, betraying their age; others of a paler hue, showing that the sun had not yet mellowed them into a deep rich colour. And on all alike, the fringe of icicles. A wonderful white world.

It was a long time before Bernardine even wished to speak. This beautiful whiteness may become monotonous after a time, but there is something very awe-inspiring about it, something which catches the soul and holds it.

The Disagreeable Man sat quietly by her side. Once or twice he bent forward to protect the camera when the sledge gave a lurch.

After some time they met a procession of sledges laden with timber; and August, the driver, and Robert Allitsen exchanged some fun and merriment with the drivers in their quaint blue smocks. The noise of the conversation, and the excitement of getting past the sledges, brought Bernardine back to speech again.

"I have never before enjoyed anything so much," she said.

"So you have found your tongue," he said. "Do you mind talking a little now? I feel rather lonely."

This was said in such a pathetic, aggrieved tone, that Bernardine laughed and looked at her companion. His face wore an unusually bright expression. He was evidently out to enjoy himself.

"*You* talk," she said; "and tell me all about the country."

And he told her what he knew, and, amongst

other things, about the avalanches. He was able to point out where some had fallen the previous year. He stopped in the middle of his conversation to tell her to put up her umbrella.

"I can't trouble to hold it for you," he said; "but I don't mind opening it. The sun is blazing to-day, and you will get your eyes bad if you are not careful. That would be a pity, for you seem to me rather better lately."

"What a confession for you to make of any one!" said she.

"Oh, I don't mean to say that you will ever get well," he added grimly. "You seem to have pulled yourself in too many directions for that. You have tried to be too alive; and now you are obliged to join the genus cabbage."

"I am certainly less ill than I was when I first came," she said; "and I feel in a better frame of mind altogether. I am learning a good deal in sad Petershof."

"That is more than I have done," he answered.

"Well, perhaps you teach instead," she said. "You have taught me several things. Now, go on telling me about the country people. You like them?"

"I love them," he said simply. "I know them well, and they know me. You see I have been in this district so long now, and have walked about so much, that the very woodcutters know me; and the drivers give me lifts on their piles of timber."

"You are not surly with the poor people, then?" said Bernardine; "though I must say I cannot imagine you being genial. Were you ever genial, I wonder?"

"I don't think that has ever been laid to my charge," he answered.

The time passed away pleasantly. The Disagreeable Man was scarcely himself to-day; or was it that he was more like himself? He seemed in a boyish mood; he made fun out of nothing, and laughed with such young fresh laughter, that even August, the grave blue-spectacled driver, was moved to mirth. As for Bernardine, she had to look at Robert Allitsen several times to be sure that he was the same Robert Allitsen she had known two hours ago in Petershof. But she made no remark, and showed no surprise, but met his merriness half way. No one could be a cheerier companion than herself when she chose.

At last they arrived at Loschwitz. The sledge wound its way through the sloshy streets of the queer little village, and finally drew up in front of the Gasthaus. It was a black sunburnt châlet, with green shutters, and steps leading up to a green balcony. A fringe of sausages hung from the roof; red bedding was scorching in the sunshine; three cats were sunning themselves on the steps; a young woman sat in the green balcony knitting. There were some curious inscriptions on the walls of the châlet, and the date was distinctly marked, "1670."

An old woman over the way sat in her doorway spinning. She looked up as the sledge stopped before the Gasthaus; but the young woman in the green balcony went on knitting, and saw nothing.

A buxom elderly Hausfrau came out to greet the guests. She wore a naturally kind expression on her old face, but when she saw who the gentleman was, the kindness positive increased to kindness superlative.

She first retired and called out:

"Liza, Fritz, Liza, Trüdchen, come quickly!"

Then she came back, and cried:

"Herr Allitsen, what a surprise!"

She shook his hand times without number, greeted Bernardine with motherly tenderness, and interspersed all her remarks with frantic cries of "Liza, Fritz, Trüdchen, make haste!"

She became very hot and excited, and gesticulated violently.

All this time the young woman sat knitting, but not looking up. She had been beautiful, but her face was worn now, and her eyes had that vacant stare which betokened the vacant mind.

The mother whispered to Robert Allitsen:

"She notices no one now; she sits there always waiting."

Tears came into the kind old eyes.

Robert Allitsen went and bent down to the young woman, and held out his hand.

"Catharina," he said gently.

She looked up then, and saw him, and recognized him.

Then the sad face smiled a welcome.

He sat near her, and took her knitting in his hand, pretending to examine what she had done, chatting to her quietly all the time. He asked her what she had been doing with herself since he had last seen her, and she said:

"Waiting. I am always waiting."

He knew that she referred to her lover, who had been lost in an avalanche the eve before their wedding morning. That was four years ago, but Catharina was still waiting. Allitsen remembered her as a bright young girl, singing in the Gasthaus, waiting cheerfully on the guests: a bright gracious presence. No one could cook trout as she could; many a dish of trout had she served up for him. And now she sat in the sunshine knitting and waiting, scarcely ever looking up. That was her life.

"Catharina," he said, as he gave her back her knitting, "do you remember how you used to cook me the trout?"

Another smile passed over her face. Yes, she remembered.

"Will you cook me some to-day?"

She shook her head, and returned to her knitting.

Bernardine watched the Disagreeable Man with amazement. She could not have believed that his manner could be so tender and kindly. The old mother standing near her whispered:

"He was always so good to us all; we love him, every one of us. When poor Catharina was betrothed five years ago, it was to Herr Allitsen we first told the good news. He has a wonderful way about him—just look at him with Catharina now. She has not noticed any one for months, but she knows him, you see."

At that moment the other members of the household came: Liza, Fritz, and Trüdchen; Liza, a maiden of nineteen, of the homely Swiss type; Fritz, a handsome lad of fourteen; and Trüdchen, just free from school, with her school-satchel swung on her back. There was no shyness in their greeting; the Disagreeable Man was evidently an old and much-loved friend, and inspired confidence, not awe. Trüdchen fumbled in his coat pocket, and found what she expected to find there, some sweets, which she immediately began to eat, perfectly contented and self-satisfied. She smiled and nodded at Robert Allitsen, as though to reassure him that the sweets were not bad, and that she was enjoying them.

"Liza will see to lunch," said the old mother. "You shall have some mutton cutlets and some *forellen*. But before she goes, she has something to tell you."

"I am betrothed to Hans," Liza said, blushing.

"I always knew you were fond of Hans," said the Disagreeable Man. "He is a good fellow, Liza, and I'm glad you love him. But haven't you just teased him!"

"That was good for him," Liza said brightly.

"Is he here to-day?" Robert Allitsen asked. Liza nodded.

"Then I shall take your photographs," he said.

While they had been speaking, Catharina rose from her seat, and passed into the house.

Her mother followed her, and watched her go into the kitchen.

"I should like to cook the *forellen*," she said very quietly.

It was months since she had done anything in the house. The old mother's heart beat with pleasure.

"Catharina, my best loved child!" she whispered; and she gathered the poor suffering soul near to her.

In about half an hour the Disagreeable Man and Bernardine sat down to their meal. Robert Allitsen had ordered a bottle of Sassella, and he was just pouring it out when Catharina brought in the *forellen*.

"Why, Catharina," he said, "you don't mean you've cooked them? Then they will be good!"

She smiled, and seemed pleased, and then went out of the room.

Then he told Bernardine her history, and spoke with such kindness and sympathy that Bernardine was again amazed at him. But she made no remark.

"Catharina was always sorry that I was ill," he

G

said. "When I stayed here, as I have done, for weeks together, she used to take every care of me. And it was a kindly sympathy which I could not resent. In those days I was suffering more than I have done for a long time now, and she was very pitiful. She could not bear to hear me cough. I used to tell her that she must learn not to feel. But you see she did not learn her lesson, for when this trouble came on her, she felt too much. And you see what she is."

They had a cheery meal together, and then Bernardine talked with the old mother, whilst the Disagreeable Man busied himself with his camera. Liza was for putting on her best dress, and doing her hair in some wonderful way. But he would not hear of such a thing. But seeing that she looked disappointed, he gave in, and said she should be photographed just as she wished; and off she ran to change her attire. She went up to her room a picturesque, homely working girl, and she came down a tidy, awkward-looking young woman, with all her finery on, and all her charm off.

The Disagreeable Man grunted, but said nothing.

Then Hans arrived, and then came the posing, which caused much amusement. They both stood perfectly straight, just as a soldier stands before presenting arms. Both faces were perfectly expressionless. The Disagreeable Man was in despair.

"Look happy!" he entreated.

They tried to smile, but the anxiety to do so produced an expression of melancholy which was too much for the gravity of the photographer. He laughed heartily.

"Look as though you weren't going to be photographed," he suggested. "Liza, for goodness' sake look as though you were baking the bread; and Hans, try and believe that you are doing some of your beautiful carving."

The patience of the photographer was something wonderful. At last he succeeded in making them appear at their ease. And then he told Liza that she must go and change her dress, and be photographed now in the way he wished. She came down again, looking fifty times prettier in her working clothes.

Now he was in his element. He arranged Liza and Hans on the sledge of timber, which had then driven up, and made a picturesque group of them all: Hans and Liza sitting side by side on the timber, the horses standing there so patiently after their long journey through the forests, the driver leaning against his sledge smoking his long china pipe.

"That will be something like a picture," he said to Bernardine, when the performance was over. "Now I am going for about a mile's walk. Will you come with me and see what I am going to photograph, or will you rest here till I come back?"

She chose the latter, and during his absence was shown the treasures and possessions of a Swiss peasant's home.

She was taken to see the cows in the stalls, and had a lecture given her on the respective merits of Schneewitchen, a white cow, Kartoffelkuchen, a dark brown one, and Röslein, the beauty of them all. Then she looked at the spinning-wheel, and watched the old Hausfrau turn the treadle. And so the time passed, Bernardine making good friends of them all. Catharina had returned to her knitting, and began working, and, as before, not noticing any one. But Bernardine sat by her side, playing with the cat, and after a time Catharina looked up at Bernardine's little thin face, and, after some hesitation, stroked it gently with her hand.

"Fräulein is not strong," she said tenderly, "If Fräulein lived here, I should take care of her."

That was a remnant of Catharina's past She had always loved everything that was ailing and weakly.

Her hand rested on Bernardine's hand. Bernardine pressed it in kindly sympathy, thinking the while of the girl's past happiness and present bereavement.

"Liza is betrothed," she said, as though to herself. "They don't tell me; but I know. I was betrothed once."

She went on knitting. And that was all she said of herself.

Then after a pause she said:

"Fräulein is betrothed?"

Bernardine smiled, and shook her head, and Catharina made no further inquiries. But she looked up from her work from time to time, and seemed pleased that Bernardine still stayed with her. At last the old mother came to say that the coffee was ready, and Bernardine followed her into the parlour.

She watched Bernardine drinking the coffee, and finally poured herself out a cup too.

"This is the first time Herr Allitsen has ever brought a friend," she said. "He has always been alone. Fräulein is betrothed to Herr Allitsen—is that so? Ah, I am glad. He is so good and so kind."

Bernardine stopped drinking her coffee.

"No, I am not betrothed," she said cheerily. "We are just friends; and not always that either. We quarrel."

"All lovers do that," persisted Frau Steinhart triumphantly.

"Well, you ask him yourself," said Bernardine, much amused. She had never looked upon Robert Allitsen in that light before. "See, there he comes."

Bernardine was not present at the court martial, but this was what occurred. Whilst the Disagreeable Man was paying the reckoning, Frau Steinhart said in her most motherly tones:

"Fräulein is a very dear young lady: Herr Allitsen has made a wise choice. He is betrothed at last."

The Disagreeable Man stopped counting out the money.

"Stupid old Frau Steinhart!" he said good-naturedly. "People like myself don't get betrothed. We get buried instead!"

"Na, na!" she answered. "What a thing to say—and so unlike you too! No, but tell me."

"Well, I am telling you the truth," he replied. "If you won't believe me, ask Fräulein herself."

"I have asked her," said Frau Steinhart, "and she told me to ask you."

The Disagreeable Man was much amused. He had never thought of Bernardine in that way.

He paid the bill, and then did something which rather astonished Frau Steinhart, and half convinced her.

He took the bill to Bernardine, told her the amount of her share, and she repaid him then and there.

There was a twinkle in her eye as she looked up at him. Then the composure of her features relaxed, and she laughed.

He laughed too, but no comment was made upon the episode. Then began the good-byes, and the preparations for the return journey.

Bernardine bent over Catharina, and kissed her sad face.

"Fräulein will come again?" she whispered eagerly.

And Bernardine promised. There was something in Bernardine's manner which had won the poor girl's fancy: some unspoken sympathy, some quiet geniality.

Just as they were starting, Frau Steinhart whispered to Robert Allitsen:

"It is a little disappointing to me, Herr Allitsen. I did so hope you were betrothed."

August, the blue-spectacled driver, cracked his whip, and off the horses started homewards.

For some time there was no conversation between the two occupants of the sledge. Bernardine was busy thinking about the experiences of the day, and the Disagreeable Man seemed in a brown study. At last he broke the silence by asking her how she liked his friends, and what she thought of Swiss home life; and so the time passed pleasantly.

He looked at her once, and said she seemed cold.

"You are not warmly clothed," he said. "I have an extra coat. Put it on; don't make a fuss, but do so at once. I know the climate, and you don't."

She obeyed, and said she was all the cosier for it.

As they were nearing Petershof, he said half-nervously:

"So my friends took you for my betrothed. I hope you are not offended."

"Why should I be?" she said frankly. "I was only amused, because there never were two people less lover-like than you and I are."

"No, that's quite true," he replied, in a tone of voice which betokened relief.

"So that I really don't see that we need concern ourselves further in the matter," she added, wishing to put him quite at his ease. "I'm not offended, and you are not offended, and there's an end of it."

"You seem to me to be a very sensible young woman in some respects," the Disagreeable Man remarked after a pause. He was now quite cheerful again, and felt he could really praise his companion. "Although you have read so much, you seem to me sometimes to take a sensible view of things. Now, I don't want to be betrothed to you, any more than I suppose you want to be betrothed to me. And yet we can talk quietly about the matter without a scene. That would be impossible with most women."

Bernardine laughed.

"Well, I only know," she said cheerily, "that I have enjoyed my day very much, and I'm much obliged to you for your companionship. The fresh air, and the change of surroundings, will have done me good."

His reply was characteristic of him.

"It is the least disagreeable day I have spent for many months," he said quietly.

"Let me settle with you for the sledge now," she said, drawing out her purse, just as they came in sight of the Kurhaus.

They settled money matters, and were quits.

Then he helped her out of the sledge, and he stooped to pick up the shawl she dropped.

"Here is the shawl you are always dropping," he said. "You're rather cold, aren't you? Here, come to the restaurant and have some brandy. Don't make a fuss. I know what's the right thing for you."

She followed him to the restaurant, touched by his rough kindness. He himself took nothing, but he paid for her brandy.

That evening after *table-d'hôte*, or rather after he had finished his dinner, he rose to go to his room as usual. He generally went off without a remark. But to-night he said:

"Good-night, and thank you for your companionship. It has been my birthday to-day, and I've quite enjoyed it."

CHAPTER XI.

IF ONE HAS MADE THE ONE GREAT SACRIFICE."

THERE was a suicide in the Kurhaus one afternoon. A Dutchman, Vandervelt, had received rather a bad account of himself from the doctor a few days previously, and in a fit of depression, so it was thought, he had put a bullet through his head. It had occurred through Marie's unconscious agency. She found him lying on his sofa when she went as usual to take him his afternoon glass of milk. He asked her to give him a packet which was on the top shelf of his cupboard.

"Willingly," she said, and she jumped nimbly on the chair, and gave him the case.

"Anything more?" she asked kindly, as she

watched him draw himself up from the sofa. She thought at the time that he looked wild and strange; but then, as she pathetically said afterwards, who did not look wild and strange in the Kurhaus?

"Yes," he said. "Here are five francs for you."

She thought that rather unusual too; but five francs, especially coming unexpectedly like that, were not to be despised, and Marie determined to send them off to that *Mutterli* at home in the nut-brown châlet at Grüsch.

So she thanked Mynheer van Vandervelt, and went off to her pantry to drink some cold tea which the English people had left, and to clean the lamps. Having done that, and knowing that the matron was busily engaged carrying on a flirtation with a young Frenchman, Marie took out her writing materials, and began a letter to her old mother. These peasants know how to love each other, and some of them know how to tell each other too. Marie knew. And she told her mother of the gifts she was bringing home, the little nothings given her by the guests.

She was very happy writing this letter: the little nut-brown home rose before her.

"Ach!" she said, "how I long to be home!"

And then she put down her pen, and sighed.

"Ach!" she said, "and when I'm there, I shall long to be here. *Da wo ich nicht bin, da ist das Glück.*"

Marie was something of a philosopher.

Suddenly she heard the report of a pistol, followed by a second report. She dashed out of her little pantry, and ran in the direction of the sound. She saw Wärli in the passage. He was looking scared, and his letters had fallen to the ground. He pointed to No. 54.

It was the Dutchman's room.

Help arrived. The door was forced open, and Vandervelt was found dead. The case from which he had taken the pistol was lying on the sofa. When Marie saw that, she knew that she had been an unconscious accomplice. Her tender heart overflowed with grief.

Whilst others were lifting him up, she leaned her head against the wall, and sobbed.

"It was my fault, it was my fault!" she cried. "I gave him the case. But how was I to know?"

They took her away, and tried to comfort her, but it was all in vain.

"And he gave me five francs," she sobbed. "I shudder to think of them."

It was all in vain that Würli gave her a letter for which she had been longing for many days.

"It is from your *Mutterli*," he said, as he put it into her hands. "I give it willingly. I don't like the look of one or two of the letters I have to give you, Mariechen. That Hans writes to you. Confound him!"

But nothing could cheer her. Würli went away shaking his curly head sadly, shocked at the death of the Dutchman, and shocked at Marie's sorrow. And the cheery little postman did not do much whistling that evening.

Bernardine heard of Marie's trouble, and rang for her to come. Marie answered the bell, looking the picture of misery. Her kind face was tear-stained, and her only voice was a sob.

Bernardine drew the girl to her.

"Poor old Marie," she whispered. "Come and cry your kind heart out, and then you will feel better. Sit by me here, and don't try to speak. And I will make you some tea in true English fashion, and you must take it hot, and it will do you good."

The simple sisterly kindness and silent sympathy soothed Marie after a time. The sobs ceased, and the tears also. And Marie put her hand in her pocket and gave Bernardine the five francs.

"Fräulein Holme, I hate them," she said. "I could never keep them. How could I send them now to my old mother? They would bring her ill luck—indeed they would."

The matter was solved by Bernardine in a masterly fashion. She suggested that Marie should buy flowers with the money, and put them on the Dutchman's coffin. This idea comforted Marie beyond Bernardine's most sanguine expectations.

II

"A beautiful tin wreath," she said several times. "I know the exact kind. When my father died, we put one on his grave."

That same evening, during *table-d'hôte*, Bernardine told the Disagreeable Man the history of the afternoon. He had been developing photographs, and had heard nothing. He seemed very little interested in her relation of the suicide, and merely remarked:

"Well, there's one person less in the world."

"I think you make these remarks from habit," Bernardine said quietly, and she went on with her dinner, attempting no further conversation with him. She herself had been much moved by the sad occurrence; every one in the Kurhaus was more or less upset; and there was a thoughtful, anxious expression on more than one ordinarily thoughtless face. The little French danseuse was quiet: the Portuguese ladies were decidedly tearful: the vulgar German Baroness was quite depressed: the comedian at the Belgian table ate his dinner in silence. In fact, there was a weight pressing down on all. Was it really possible, thought Bernardine, that Robert Allitsen was the only one there unconcerned and unmoved? She had seen him in a different light amongst his friends, the country folk, but it was just a glimpse which had not lasted long. The young-heartedness, the geniality, the sympathy which had so astonished her during their day's outing, astonished her still more by their total disappearance. The gruffness had returned: or had it never been absent? The lovelessness and leadenness of his temperament had once more asserted themselves: or was it that they had never for one single day been in the background?

These thoughts passed through her mind as he sat next to her reading his paper—that paper which he never passed on to any one. She hardened her heart against him; there was no need for ill-health and disappointment to have brought any one to a miserable state of indifference like that. Then she looked at his wan face and frail form, and her heart softened at once. At the moment when her heart softened to him, he astonished her by handing her his paper.

"Here is something to interest you," he said, "an article on Realism in Fiction, or some nonsense like that. You needn't read it now. I don't want the paper again."

"I thought you never lent anything," she said, as she glanced at the article, "much less gave it."

"Giving and lending are not usually in my line," he replied. "I think I told you once that I thought selfishness perfectly desirable and legitimate, if one had made the one great sacrifice."

"Yes," she said eagerly; "I have often wondered what you considered the one great sacrifice."

"Come out into the air," he answered, "and I will tell you."

She went to put on her cloak and hat, and found him waiting for her at the top of the staircase. They passed out into the beautiful night: the sky was radiantly bejewelled, the air crisp and cold, and harmless to do ill. In the distance, the jodelling of some peasants. In the hotels, the fun and merriment, side by side with the suffering and hopelessness. In the deaconess's house, the body of the Dutchman. In God's heavens, God's stars.

Robert Allitsen and Bernardine walked silently for some time.

"Well," she said, "now tell me."

"The one great sacrifice," he said half to himself, "is the going on living one's life for the sake of another, when everything that would seem to make life acceptable has been wrenched away, not the pleasures, but the duties, and the possibilities of expressing one's energies, either in one direction or another: when, in fact, living is only a long tedious dying. If one has made this sacrifice, everything else may be forgiven."

He paused a moment, and then continued:

"I have made this sacrifice, therefore I consider I have done my part without flinching. The greatest thing I had to give up, I gave up: my death. More could not be required of any one."

He paused again, and Bernardine was silent from mere awe.

"But freedom comes at last," he said, "and some day I shall be free. When my mother dies, I shall be free. She is old. If I were to die, I should break her heart, or rather she would fancy that her heart was broken. (And it comes to the same thing). And I should not like to give her more grief than she has had. So I am just waiting. It may be months, or weeks, or years. But I know how to wait: if I have not learnt anything else, I have learnt how to wait. And then"......

Bernardine had unconsciously put her hand on his arm; her face was full of suffering.

"And then?" she asked, with almost painful eagerness.

"And then I shall follow your Dutchman's example," he said deliberately.

Bernardine's hand fell from the Disagreeable Man's arm.

She shivered.

"You are cold, you little thing," he said, almost tenderly for him. "You are shivering."

"Was I?" she said, with a short laugh. "I was wondering when you would get your freedom, and whether you would use it in the fashion you now intend."

"Why should there be any doubt?" he asked.

"One always hopes there would be a doubt," she said, half in a whisper.

Then he looked up, and saw all the pain on the little face.

CHAPTER XII.

THE DISAGREEABLE MAN MAKES A LOAN.

The Dutchman was buried in the little cemetery which faced the hospital. Marie's tin wreath was placed on the grave. And there the matter ended. The Kurhaus guests recovered from their depression: the German Baroness returned to her buoyant vulgarity, the little danseuse to her busy flirtations. The French Marchioness, celebrated in Parisian circles for her domestic virtues, from which she was now taking a holiday, and a very considerable holiday too, gathered her nerves together again and took renewed pleasure in the society of the Russian gentleman. The French Marchioness had already been requested to leave three other hotels in Petershof; but it was not at all probable that the proprietors of the Kurhaus would have presumed to measure Madame's morality or immorality. The Kurhaus committee had a benign indulgence for humanity— provided of course that humanity had a purse —an indulgence which some of the English hotels would not have done badly to imitate. There was a story afloat concerning the English quarter, that a tired little English lady, of no importance to look at, probably not rich, and probably not handsome, came to the most respectable hotel in Petershof, thinking to find there the peace and quiet which her weariness required.

But no one knew who the little lady was, whence she had come, and why. She kept entirely to herself, and was thankful for the luxury of loneliness after some overwhelming sorrow.

One day she was requested to go. The proprietor of the hotel was distressed, but he could not do otherwise than comply with the demands of his guests.

"It is not known who you are, Mademoiselle," he said. "And you are not approved of. You English are curious people. But what can I do? You have a cheap room, and are a stranger to me. The others have expensive apartments, and come year after year. You see my position, Mademoiselle? I am sorry."

So the little tired lady had to go. That was how the story went. It was not known what became of her, but it was known that the English people in the Kurhaus tried to persuade her to come to them. But she had lost heart, and left in distress.

This could not have happened in the Kurhaus, where all were received on equal terms, those about whom nothing was known, and those about whom too much was known. The strange mixture and the contrasts of character afforded endless scope for observation and amusement, and Bernardine, who was daily becoming more interested in her surroundings, felt that she would have been sorry to have exchanged her present abode for the English quarter. The amusing part of it was that the English people in the Kurhaus were regarded by their compatriots in the English quarter as sheep of the blackest dye! This was all the more ridiculous because with two exceptions—firstly of Mrs. Reffold, who took nearly all her pleasures with the American colony in the Grand Hotel; and secondly, of a Scotch widow who had returned to Petershof to weep over her husband's grave, but put away her grief together with her widow's weeds, and consoled herself with a Spanish gentleman—with these two exceptions, the little English community in the Kurhaus was most humdrum and harmless, being occupied, as in the case of the Disagreeable Man, with cameras and cheese-mites, or in other cases with the still more engrossing pastime of taking care of one's ill-health, whether real or fancied: but yet, an innocent hobby in itself, and giving one absolutely no leisure to do anything worse: a great recommendation for any pastime.

This was not Bernardine's occupation: it was difficult to say what she did with herself, for she had not yet followed Robert Allitsen's advice and taken up some definite work; and the very fact that she had no such wish, pointed probably to a state of health which forbade it. She, naturally so keen and hard-working, was content to take what the hour brought, and the hour brought various things: chess with the Swedish professor, or Russian dominoes with the shrivelled-up little Polish governess who always tried to cheat, and who clutched her tiny winnings with precisely the same greediness shown by the Monte Carlo female gamblers. Or the hour brought a stroll with the French danseuse and her poodle, and a conversation about the mere trivialities of life, which a year or two, or even a few months ago, Bernardine would have condemned as beneath contempt, but which were now taking their rightful place in her new standard of importances. For some natures learn with greater difficulty and after greater delay than others, that the real importances of our existence are the nothingnesses of every-day life, the nothingnesses which the philosopher in his study, reasoning about and analysing human character, is apt to overlook; but which, nevertheless, make him and every one else more of a human reality and less of an abstraction. And Bernardine, hitherto occupied with so-called intellectual pursuits, with problems of the study, of no value to the great world outside the study, or with social problems of the great world, great movements, and great questions, was now just beginning to appreciate the value of the little incidents of that same great world. Or the hour brought its own thoughts, and Bernardine found herself constantly thinking of the Disagreeable Man: always in sorrow and always with sympathy, and sometimes with tenderness.

When he told her about the one sacrifice, she could have wished to wrap him round with love and tenderness. If he could only have known it, he had never been so near love as then. She had suffered so much herself, and, with increasing weaknesses, had so wished to put off the burden of the flesh, that her whole heart went out to him.

Would he get his freedom, she wondered, and would he use it? Sometimes when she was with him, she would look up to see whether she could read the answer in his face; but she never saw any variation of expression there, nothing to give her even a suggestion. But this she noticed: that there was a marked variation in his manner, and that when he had been rough in bearing, or bitter in speech, he made silent amends at the earliest opportunity by being less rough and less bitter. She felt this was no small concession on the part of the Disagreeable Man.

He was particularly disagreeable on the day when the Dutchman was buried, and so the following day when Bernardine met him in the little English library, she was not surprised to find him almost kindly.

He had chosen the book which she wanted, but he gave it up to her at once without any grumbling, though Bernardine expected him to change his mind before they left the library.

"Well," he said, as they walked along together, "and have you recovered from the death of the Dutchman?"

"Have you recovered, rather let me ask?"

she said. "You were in a horrid mood last night."

"I was feeling wretchedly ill," he said quietly.

That was the first time he had ever alluded to his own health.

"Not that there is any need to make an excuse," he continued, "for I do not recognise that there is any necessity to consult one's surroundings, and alter the inclination of one's mind accordingly. Still, as a matter of fact, I felt very ill."

"And to-day?" she asked.

"To-day I am myself again," he answered quickly: "that usual normal self of mine, whatever that may mean. I slept well, and I dreamed of you. I can't say that I had been thinking of you, because I had not. But I dreamed that we were children together, and playmates. Now that was very odd: because I was a lonely child, and never had any playmates."

"And I was lonely too," said Bernardine.

"Every one is lonely," he said, "but every one does not know it.'

"But now and again the knowledge comes like a revelation," she said, "and we realise that we stand practically alone, out of any one's reach for help or comfort. When you come to think of it, too, how little able we are to explain ourselves. When you have wanted to say something which was burning within you, have you not noticed on the face of the listener that unmistakable look of non-comprehension, which throws you back on yourself? That is one of the moments when the soul knows its own loneliness."

Robert Allitsen looked up at her.

"You little thing," he said, "you put things neatly sometimes. "You have felt, haven't you?"

"I suppose so," she said. "But that is true of most people."

"I beg your pardon," he answered, "most people neither think nor feel: unless they think they have an ache, and then they feel it!"

"I believe," said Bernardine, "that there is more thinking and feeling than one generally supposes."

"Well, I can't be bothered with that now," he said. "And you interrupted me about my dream. That is an annoying habit you have."

"Go on," she said. "I apologize."

"I dreamed we were children together, and playmates," he continued. "We were not at all happy together, but still we were playmates. There was nothing we did not quarrel about. You were disagreeable, and I was spiteful. Our greatest dispute was over a Christmas-tree. And that was odd, too, for I have never seen a Christmas-tree."

"Well?" she said, for he had paused. "What a long time you take to tell a story."

"You were not called Bernardine," he said. "You were called by some ordinary sensible name. I don't remember what. But you were very disagreeable. That I remember well. At last you disappeared, and I went about looking for you. 'If I can find something to cause a quarrel,' I said to myself, ' she will come back.' So I went and smashed your doll's head. But you did not come back. Then I set on fire your doll's house. But even that did not bring you back. Nothing brought you back. That was my dream. I hope you are not offended. Not that it makes any difference if you are."

Bernardine laughed.

"I am sorry that I should have been such an unpleasant playmate," she said. "It was a good thing I did disappear."

"Perhaps it was," he said. "There would have been a terrible scene about that doll's head. An odd thing for me to dream about Christmas-trees and dolls and playmates: especially when I went to sleep thinking about my new camera."

"You have a new camera?" she asked.

"Yes," he answered, "and a beauty, too. Would you like to see it?"

She expressed a wish to see it, and when they reached the Kurhaus, she went with him up to his beautiful room, where he spent his time in the company of his microscope and his chemical bottles and his photographic possessions.

"If you sit down and look at those photographs, I will make you some tea," he said

"There is the camera, but please not to touch it until I am ready to show it myself."

She watched him preparing the tea; he did everything so daintily, this Disagreeable Man. He put a handkerchief on the table, to serve for an afternoon tea-cloth, and a tiny vase of violets formed the centre-piece. He had no cups, but he polished up two tumblers, and no housemaid could have been more particular about their glossiness. Then he boiled the water and made the tea. Once she offered to help him; but he shook his head.

"Kindly not to interfere," he said grimly. "No one can make tea better than I can."

After tea, they began the inspection of the new camera, and Robert Allitsen showed her all the newest improvements. He did not seem to think much of her intelligence, for he explained everything as though he were talking to a child, until Bernardine rather lost patience.

"You need not enter into such elaborate explanations," she suggested. "I have a small amount of intelligence, though you do not seem to detect it."

He looked at her as one might look at an impatient child.

"Kindly not to interrupt me," he replied mildly. "How very impatient you are! And how restless! What must you have been like before you fell ill?"

But he took the hint all the same, and shortened his explanations, and as Bernardine was genuinely interested, he was well satisfied. From time to time he looked at his old camera and at his companion, and from the expression of unease on his face, it was evident that some contest was going on in his mind. Twice he stood near his old camera, and turned round to Bernardine intending to make some remark. Then he changed his mind, and walked abruptly to the other end of the room as though to seek advice from his chemical bottles. Bernardine meanwhile had risen from her chair, and was looking out of the window.

"You have a lovely view," she said. "It must be nice to look at that when you are tired of dissecting cheese-mites. All the same,

I think the white scenery gives one a great sense of sadness and loneliness."

"Why do you speak always of loneliness?" he asked.

"I have been thinking a good deal about it," she said. "When I was strong and vigorous, the idea of loneliness never entered my mind. Now I see how lonely most people are. If I believed in God as a Personal God, I should be inclined to think that loneliness were part of his scheme: so that the soul of man might turn to him and him alone."

The Disagreeable Man was standing by his camera again: his decision was made.

"Don't think about those questions," he said kindly. "Don't worry and fret too much about the philosophy of life. Leave philosophy alone, and take to photography instead. Here, I will lend you my old camera."

"Do you mean that?" she asked, glancing at him in astonishment.

"Of course I mean it," he said.

He looked remarkably pleased with himself, and Bernardine could not help smiling.

He looked just as a child looks when he has given up a toy to another child, and is conscious that he has behaved himself rather well.

"I am very much obliged to you," she said frankly. "I have had a great wish to learn photography."

"I might have lent my camera to you before, mightn't I?" he said thoughtfully.

"No," she answered. "There was not any reason."

"No," he said, with a kind of relief, "there was not any reason. That is quite true."

"When will you give me my first lesson?" she asked. "Perhaps, though, you would like to wait a few days, in case you change your mind."

"It takes me some time to make up my mind," he replied; "but I do not change it. So I will give you your first lesson to-morrow. Only you must not be impatient. You must consent to be taught; you cannot possibly know everything!"

They fixed a time for the morrow, and

Bernardine went off with the camera; and meeting Marie on the staircase, confided to her the piece of good fortune which had befallen her.

"See what Herr Allitsen has lent me, Marie!" she said.

Marie raised her hands in astonishment.

'Who would have thought such a thing of Herr Allitsen?" said Marie. "Why, he does not like lending me a match."

Bernardine laughed and passed on to her room.

And the Disagreeable Man meanwhile was cutting a new scientific book which had just come from England. He spent a good deal of money on himself. He was soon absorbed in this book, and much interested in the diagrams.

Suddenly he looked up to the corner where the old camera had stood, before Bernardine took it away in triumph.

"I hope she won't hurt that camera," he said a little uneasily. "I am half sorry that" . . . .

Then a kinder mood took possession of him.

"Well, at least it will keep her from fussing and fretting and thinking. Still, I hope she won't hurt it."

CHAPTER XIII.

A DOMESTIC SCENE.

ONE afternoon when Mrs. Reffold came to say good-bye to her husband before going out for the usual sledge-drive, he surprised her by his unwonted manner.

"Take your cloak off," he said sharply. "You cannot go for your drive this afternoon. You don't often give up your time to me; you must do so to-day."

She was so astonished, that she at once laid aside her cloak and hat, and touched the bell.

"Why are you ringing?" Mr. Reffold asked testily.

"To send a message of excuse," she answered, with provoking cheerfulness.

She scribbled something on a card, and gave it to the servant who answered the bell.

"Now," she said, with great sweetness of manner. And she sat down beside him, drew out her fancy-work, and worked away contentedly. She would have made a charming study of a devoted wife soothing a much-loved husband in his hours of sickness and weariness.

"Do you mind giving up your drive?" he asked.

"Not in the least," she replied. "I am rather tired of sledging."

"You soon get tired of things, Winifred," he said.

"Yes, I do," was the answer. "I am so easily bored. I am quite tired of this place."

"You will have to stay here a little longer," he said, "and then you will be free to go where you choose. I wish I could die quicker for you, Winifred."

Mrs. Reffold looked up from her embroidery.

"You will get better soon," she said. "You are better."

"Yes, you've helped a good deal to make me

better," he said bitterly. "You have been a most unselfish person, haven't you? You have given me every care and attention, haven't you?"

"You seem to me in a very strange mood to-day," she said, looking puzzled. "I don't understand you."

Mr. Reffold laughed.

"Poor Winifred," he said. "If it is ever your lot to fall ill and be neglected, perhaps then you will think of me."

"Neglected?" she said, in some surprise. "What do you mean? I thought you had everything you wanted. The nurse brought excellent testimonials. I was careful in the choice of her. You have never complained before."

He turned wearily on his side, and made no answer. And for some time there was silence between them. Then he watched her as she bent over her embroidery.

"You are very beautiful, Winifred," he said quietly, "but you are a selfish woman. Has it ever struck you that you are selfish?"

Mrs. Reffold gave no reply, but she made a resolution to write to her particular friend at Cannes and confide to her how very trying her husband had become.

"I suppose it is part of his illness," she thought meekly. "But it is hard to have to bear it."

And Mrs. Reffold pitied herself profoundly. She stitched sincere pity for herself into that piece of embroidery.

"I remember you telling me," continued Mr. Reffold, " that sick people repelled you. That was when I was strong and vigorous. But since I have been ill, I have often recalled your words. Poor Winifred! You did not think then that you would have an invalid husband on your hands. Well, you were not intended for sick-room nursing, and you have not tried to be what you were not intended for. Perhaps you were right, after all."

"I don't know why you should be so unkind to-day," Mrs. Reffold said, with pathetic patience. "I can't understand you. You have never spoken like this before."

"No," he said; "but I have thought like this before. All the hours you have left me lonely, I have been thinking like this, with my heart full of bitterness against you, until that little girl, that Little Brick came along."

After that, it was some time before he spoke. He was thinking of his Little Brick, and of all the pleasant hours he had spent with her, and of the kind, wise words she had spoken to him, an ignorant fellow. She was something like a companion.

So he went on thinking, and Mrs. Reffold went on embroidering. She was now feeling herself to be almost a heroine. It is a very easy matter to make oneself into a heroine or a martyr. Selfish, neglectful? What did he mean? Oh, it was just part of his illness. She must go on bearing her burden as she had borne it these many months. Her rightful position was in a London ball-room. Instead of which, she had to be shut up in an Alpine village: a hard lot. It was little enough pleasure she could get, and apparently her husband grudged her that. His manner to her this afternoon was not such as to encourage her to stay in from her drive on another occasion. To-morrow she would go sledging.

That flash of light which reveals ourselves to ourselves had not yet come to Mrs. Reffold.

She looked at her husband, and thought from his restfulness that he had gone to sleep, and she was just beginning to write to that particular friend at Cannes, to tell her what a trial she was undergoing, when Mr. Reffold called her to his side.

"Winifred," he said gently, and there was tenderness in his voice, and love written on his face, "Winifred, I am sorry if I have been sharp to you. Little Brick says we mustn't come down like sledge-hammers on each other; and that is what I have been doing this afternoon. Perhaps I have been hard: I am such an illness to myself, that I must be an illness to others too. And you weren't meant for this sort of thing—were you? You are a bright beautiful creature, and I am an unfortunate dog not to have been able to make you happier. I know I am irritable. I can't help myself, indeed I can't."

This great long fellow was so yearning for love and sympathy.

What would it not have been to him if she had gathered him into her arms, and soothed all his irritability and suffering with her love?

But she pressed his hand, and kissed him lightly on the cheek, and told him that he *had* been a little sharp, but that she quite understood, and that she was not hurt. Her charm of manner gave him some satisfaction; and when Bernardine came in a few minutes later, she found Mr. Reffold looking happier and more contented than she had ever seen him. Mrs. Reffold, who was relieved at the interruption, received Bernardine warmly, though there was a certain amount of shyness which she had never been able to conquer in Bernardine's presence. There was something in the younger woman which quelled Mrs. Reffold: it may have been some mental quality, or it may have been her boots!

"Little Brick," said Mr. Reffold, "isn't it

nice to have Winifred here? And I have been so disagreeable and snappish."

"Oh, we won't say anything about that now," said Mrs. Reffold, smiling sweetly.

"But I've said I am sorry,' he continued. "And one can't do more."

"No," said Bernardine, who was amused at the notion of Mr. Reffold apologizing to Mrs. Reffold, and of Mrs. Reffold posing as the gracious forgiver, "one can't do more." But she could not control her feelings, and she laughed.

"You seem rather merry this afternoon," Mr. Reffold said, in a reproachful tone of voice.

"Yes," she said. And she laughed again. Mrs. Reffold's forgiving graciousness had altogether upset her gravity.

"You might at least tell us the joke," Mrs. Reffold said.

Bernardine looked at her hopelessly, and laughed again.

"I have been developing photographs all the afternoon," she said, "and I suppose the closeness of the air and the badness of my negatives have been too much for me. Anyway, I know I must seem very rude."

She recovered herself after that, and tried hard not to think of Mrs. Reffold as the dispenser of forgiveness, although it was some time before she could look at her hostess without wishing to laugh. The corners of her mouth twitched, and her brown eyes twinkled mischievously, and she spoke very rapidly, making fun of her first attempts at photography, and criticising herself so comically, that both Mr. and Mrs. Reffold were much amused.

All the same, Bernardine was relieved when Mrs. Reffold went to fetch some silks, and left her with Mr. Reffold.

"I am very happy this afternoon, Little Brick," he said to her. "My wife has been sitting with me. But instead of enjoying the pleasure as I ought to have done, I began to find fault with her. I don't know how long I should not have gone on grumbling, but that I suddenly recollected what you taught me: that we were not to come down like sledge-hammers on each other's failings. When I remembered that, it was quite easy to forgive all the neglect and thoughtlessness. Since you have talked to me, Little Brick, everything has become easier to me."

"It is something in your own mind which has worked this," she said; "your own kind, generous mind, and you put it down to my words."

But he shook his head.

"If I knew of any poor unfortunate devil that wanted to be eased and comforted," he said, "I should tell him about you, Little Brick. You have been very good to me. You may be clever, but you have never worried my stupid brain with too much scholarship. I'm just an ignorant chap, and you've never let me feel it."

He took her hand and raised it reverently to his lips.

"I say," he continued, "tell my wife it made me happy to have her with me this afternoon; then perhaps she will stay in another time. I should like her to know. And she was sweet in her manner, wasn't she? And, by Jove, she is beautiful! I am glad you have seen her here to-day. It must be dull for her with an invalid like me. And I know I am irritable. Go and tell her that she made me happy—will you?"

The little bit of happiness at which the poor fellow snatched, seemed to make him more pathetic than before. Bernardine promised to tell his wife, and went off to find her, making as an excuse a book which Mrs. Reffold had offered to lend her. Mrs. Reffold was in her bedroom. She asked Bernardine to sit down whilst she searched for the book. She had a very gracious manner when she chose.

"You are looking much better, Miss Holme," she said kindly. "I cannot help noticing your face. It looks younger and brighter. The bracing air has done you good."

"Yes, I am better," Bernardine said, rather astonished that Mrs. Reffold should have noticed her at all. "Mr. Allitsen informs me that I shall live, but never be strong. He settles every question of that sort to his own satisfaction, but not always to the satisfaction of other people!"

"He is a curious person," Mrs. Reffold said, smiling; "though I must say he is not quite as gruff as he used to be. You seem to be good friends with him."

She would have liked to say more on this subject, but experience had taught her that Bernardine was not to be trifled with.

"I don't know about being good friends," Bernardine said, "but I have a great sympathy for him. I know myself what it is to be cut off from work and active life. I have been through a misery. But mine is nothing to his."

She rose to go, but Mrs. Reffold detained her.

"Don't go yet," she said. "It is pleasant to have you."

She was leaning back in an arm-chair, playing with the fringe of an antimacassar.

"Oh, how tired I am of this horrid place!" she said suddenly. "And I have had a most wearying afternoon. Mr. Reffold seems to be more irritable every day. It is very hard that I should have to bear it."

Bernardine listened to her in astonishment.

"Yes," she added, "I am quite worn out. He never used to be so irritable. It is all very tiresome. It is quite telling on my health."

She looked the picture of health.

Bernardine gasped; and Mrs. Reffold continued:

"His grumbling this afternoon has been incessant; so much so that he himself was ashamed, and asked me to forgive him. You heard him, didn't you?"

"Yes, I heard him," Bernardine said.

"And of course I forgave him at once," Mrs. Reffold said piously. "Naturally one would do that, but the vexation remains all the same."

"Can these things be?" thought Bernardine to herself.

"He spoke in a most ridiculous way," she went on: "it certainly is not encouraging for me to spend another afternoon with him. I shall go sledging to-morrow."

"You generally do go sledging, don't you?" Bernardine asked mildly.

Mrs. Reffold looked at her suspiciously. She was never quite sure that Bernardine was not making fun of her.

"It is little enough pleasure I do have," she added, as though in self-defence. "And he seems to grudge me that too."

"I don't think he would grudge you anything," Bernardine said, with some warmth. "He loves you too much for that. You don't know how much pleasure you give him when you spare him a little of your time. He told me how happy you made him this afternoon. You could see for yourself that he was happy. Mrs. Reffold, make him happy whilst you still have him. Don't you understand that he is passing away from you—don't you understand, or is it that you *won't?* We all see it, all except you!"

She stopped suddenly, surprised at her boldness.

Mrs. Reffold was still leaning back in the arm-chair, her hands clasped together above her beautiful head. Her face was pale. She did not speak. Bernardine waited. The silence was unbroken save by the merry cries of some children tobogganing in the Kurhaus garden. The stillness grew oppressive, and Bernardine rose. She knew from the effort which those few words had cost her, how far removed she was from her old former self.

"Good-bye, Mrs. Reffold," she said nervously.

"Good-bye, Miss Holme," was the only answer.

CHAPTER XIV.

CONCERNING THE CARETAKERS.

The Doctors in Petershof always said that the caretakers of the invalids were a much greater anxiety than the invalids themselves. The invalids would either get better or die: one of two things probably. At any rate, you knew where you were with them. But not so with the caretakers: there was nothing they were not capable of doing—except taking reasonable care of their invalids! They either fussed about too much, or else they did not fuss about at all. They all began by doing the right thing: they all ended by doing the wrong. The fussy ones had fits of apathy, when the poor irritable patients seemed to get a little better; the negligent ones had paroxysms of attentiveness, when their invalids, accustomed to loneliness and neglect, seemed to become rather worse by being worried.

To remonstrate with the caretakers would have been folly: for they were well satisfied with their own methods.

To contrive their departure would have been an impossibility: for they were firmly convinced that their presence was necessary to the welfare of their charges. And then, too, judging from the way in which they managed to amuse themselves, they liked being in Petershof, though they never owned that to the invalids. On the contrary, it was the custom for the caretakers to depreciate the place, and to deplore the necessity which obliged them to continue there month after month. They were fond, too, of talking about the sacrifices which they made, and the pleasures which they willingly gave up in order to stay with their invalids. They said this in the presence of their invalids. And if the latter had told them by all means to pack up and go back to the pleasures which they had renounced, they would have been astonished at the ingratitude which could suggest the idea.

They were amusing characters, these caretakers. They were so thoroughly unconscious of their own deficiencies. They might neglect their own invalids, but they would look after other people's invalids, and play the nurse most soothingly and prettily where there was no call and no occasion. Then they would come and relate to their neglected dear ones what they had been doing for others: and the dear ones would smile quietly, and watch the buttons being stitched on for strangers, and the cornflour which they could not get nicely made for themselves, being carefully prepared for other people's neglected dear ones.

Some of the dear ones were rather bitter. But there were many of a higher order of intelligence, who seemed to realize that they had no right to be ill, and that being ill, and therefore a burden on their friends, they must make the best of everything, and be grateful for what was given them, and patient when anything was withheld. Others of a still higher order of understanding, attributed the eccentricities of the caretakers to one cause alone: the Petershof air. They knew it had the invariable effect of getting into the head, and upsetting the balance of those who drank deep of it. Therefore no one was to blame, and no one need be bitter. But these were the philosophers of the colony: a select and dainty few in any colony. But there were several rebels amongst the invalids, and they found consolation in confiding to each other their separate grievances. They generally held their conferences in the rooms known as the news-paper-rooms, where they were not likely to be interrupted by any caretakers who might have stayed at home because they were tired out.

To-day there were only a few rebels gathered together, but they were more than usually excited, because the Doctors had told several of them that their respective caretakers must be sent home.

"What must I do?" said little Mdlle.

Gerardy, wringing her hands. "The Doctor says that I must tell my sister to go home: that she only worries me, and makes me worse. He calls her a 'whirlwind.' If I won't tell her, then he will tell her, and we shall have some more scenes. Mon Dieu! and I am so tired of them. They terrify me. I would suffer anything rather than have a fresh scene. And I can't get her to do anything for me. She has no time for me. And yet she thinks she takes the greatest possible care of me, and devotes the whole day to me. Why, sometimes I never see her for hours together."

"Well, at least she does not quarrel with every one, as my mother does," said a Polish gentleman, M. Lichinsky. "Nearly every day she has a quarrel with some one or other; and then she comes to me and says she has been insulted. And others come to me mad with rage, and complain that they have been insulted by her. As though I were to blame! I tell them that now. I tell them that my mother's quarrels are not my quarrels. But one longs for peace. And the Doctor says I must have it, and that my mother must go home at once. If I tell her that, she will have a tremendous quarrel with the Doctor. As it is, he will scarcely speak to her. So you see, Mademoiselle Gerardy, that I, too, am in a bad plight. What am I to do?"

Then a young American spoke. He had been getting gradually worse since he came to Petershof, but his brother, a bright sturdy young fellow, seemed quite unconscious of the seriousness of his condition.

"And what am I to do?" he asked pathetically. "My brother does not even think I am ill. He says I am to rouse myself and come skating and tobogganing with him. Then I tell him that the Doctor says I must lie quietly in the sun. I have no one to take care of me, so I try to take a little care of myself, and then I am laughed at. It is bad enough to be ill; but it is worse when those who might help you a little, won't even believe in your illness. I wrote home once and told them; but they go by what he says; and they, too, tell me to rouse myself."

His cheeks were sunken, his eyes were leaden. There was no power in his voice, no vigour in his frame. He was just slipping quickly down the hill for want of proper care and understanding.

"I don't know whether I am much better off than you," said an English lady, Mrs. Bridgetower. "I certainly have a trained nurse to look after me, but she is altogether too much for me, and she does just as she pleases. She is always ailing, or else pretends to be; and she is always depressed. She grumbles from eight in the morning till nine at night. I have heard that she is cheerful with other people, but she never gives me the benefit of her brightness. Poor thing! She does feel the cold very much, but it is not very cheering to see her crouching near the stove, with her arms almost clasping it! When she is not talking of her own looks, all she says is: 'Oh, if I had only not come to Petershof!' or, 'Why did I ever leave that hospital in Manchester?' or, 'The cold is eating into the very marrow of my bones.' At first she used to read to me; but it was such a dismal performance that I could not bear to hear her. Why don't I send her home? Well, my husband will not hear of me being alone, and he thinks I might do worse than keep Nurse Frances. And perhaps I might."

"I would give a good deal to have a sister like pretty Fräulein Müller has," said little Fräulein Oberhof. "She came to look after me the other day when I was alone. She has the kindest way about her. But when my sister came in, she was not pleased to find Fräulein Sophie Müller with me. She does not do anything for me herself, and she does not like any one else to do anything either. Still, she is very good to other people. She comes up from the theatre sometimes at half-past nine —that is the hour when I am just sleepy—and she stamps about the room, and makes corn-flour for the old Polish lady. Then off she goes, taking with her the cornflour together with my sleep. Once I complained, but she said I was irritable. You can't think how teasing it is to hear the noise of the spoon stirring the corn-flour just when you are feeling drowsy. You say to yourself, 'Will that cornflour never be made? It seems to take centuries.'"

"One could be more patient if it were being made for oneself," said M. Lichinsky. "But at least, Fräulein, your sister does not quarrel with every one. You must be grateful for that mercy."

Even as he spoke, a stout lady thrust herself into the reading-room. She looked very hot and excited. She was M. Lichinsky's mother. She spoke with a whirlwind of Polish words. It is sometimes difficult to know when these people are angry and when they are pleased. But there was no mistake about Mme. Lichinsky. She was always angry. Her son rose from the sofa, and followed her to the door. Then he turned round to his confederates, and shrugged his shoulders.

"Another quarrel!" he said hopelessly.

CHAPTER XV.

WHICH CONTAINS NOTHING.

"You may have talent for other things," Robert Allitsen said one day to Bernardine, "but you certainly have no talent for photography. You have not made the slightest progress."

"I don't at all agree with you," Bernardine answered rather peevishly. "I think I am getting on very well."

"You are no judge," he said. "To begin with, you cannot focus properly. You have a crooked eye. I have told you that several times."

"You certainly have," she put in. "You don't let me forget that."

"Your photograph of that horrid little danseuse whom you like so much," he said, "is simply abominable. She looks like a fury. Well, she may be one for all I know, but in real life she has not the appearance of one."

"I think that is the best photograph I have done," Bernardine said, highly indignant. She could tolerate his uppishness about subjects of which she knew far more than he did; but his masterfulness about a subject of which she really knew nothing was more than she could bear with patience. He had not the tact to see that she was irritated.

"I don't know about it being the best," he said; "unless it is the best specimen of your inexperience. Looked at from that point of view, it does stand first."

She flushed crimson with temper.

"Nothing is easier than to make fun of others," she said fiercely. "It is the resource of the ignorant."

Then, after the fashion of angry women, having said her say, she stalked away. If there had been a door to bang, she would certainly

L

have banged it. However, she did what she could under the circumstances: she pushed a curtain roughly aside, and passed into the concert-room, where every night of the season's six months, a scratchy string orchestra entertained the Kurhaus guests. She left the Disagreeable Man standing in the passage.

"Dear me," he said thoughtfully. And he stroked his chin. Then he trudged slowly up to his room.

"Dear me," he said once more.

Arrived in his bedroom, he began to read. But after a few minutes he shut his book, took the lamp to the looking-glass and brushed his hair. Then he put on a black coat and a white silk tie. There was a speck of dust on the coat. He carefully removed that, and then extinguished the lamp.

On his way downstairs he met Marie, who gazed at him in astonishment. It was quite unusual for him to be seen again when he had once come up from *table-d'hôte*. She noticed the black coat and the white silk tie too, and reported on these eccentricities to her colleague Anna.

The Disagreeable Man meanwhile had reached the Concert Hall. He glanced around, and saw where Bernardine was sitting, and then chose a place in the opposite direction, quite by himself. He looked somewhat like a dog who has been well beaten. Now and again he looked up to see whether she still kept her seat. The bad music was a great irritation to him. But he stayed on heroically. There was no reason why he should stay. Gradually, too, the audience began to thin. Still he lingered, always looking like a dog in punishment.

At last Bernardine rose, and the Disagreeable Man rose too. He followed her humbly to the door. She turned and saw him.

"I am sorry I put you in a bad temper," he said. "It was stupid of me."

"I am sorry I got into a bad temper," she answered, laughing. "It was stupid of me."

"I think I have said enough to apologize," he said. "It is a process I dislike very much."

And with that he wished her good-night and went to his room.

But that was not the end of the matter, for the next day when he was taking his breakfast with her, he of his own accord returned to the subject.

"It was partly your own fault that I vexed you last night," he said. "You have never before been touchy, and so I have become accustomed to saying what I choose. And it is not in my nature to be flattering."

"That is a very truthful statement of yours," she said, as she poured out her coffee. "But I own I was touchy. And so I shall be again if you make such cutting remarks about my photographs."

"You *have* a crooked eye," he said grimly. "Look there, for instance! You have poured your coffee outside the cup. Of course you can do as you like, but the usual custom is to pour it inside the cup."

They both laughed, and the good understanding between them was cemented again.

"You are certainly getting better," he said

suddenly. "I should not be surprised if you were able to write a book after all. Not that a new book is wanted. There are too many books as it is; and not enough people to dust them. Still, it is not probable that you would be considerate enough to remember that. You will write your book."

Bernardine shook her head.

"I don't seem to care now," she said. "I think I could now be content with a quieter and more useful part."

"You will write your book," he continued. "Now listen to me. Whatever else you may do, don't make your characters hold long discussions with each other. In real life, people do not talk four pages at a time without stopping. Also, if you bring together two clever men, don't make them talk cleverly. Clever people do not. It is only the stupid ones who think they must talk cleverly all the time. And don't detain your reader too long: if you must have a sunset, let it be a short one. I could give you many more hints which would be useful to you."

"But why not use your own hints for yourself?" she suggested.

"That would be selfish of me," he said solemnly. "I wish you to profit by them."

"You are learning to be unselfish at a very rapid rate," Bernardine said.

At that moment Mrs. Reffold came into the breakfast-room, and, seeing Bernardine, gave her a stiff bow.

"I thought you and Mrs. Reffold were such friends," Robert Allitsen said.

Bernardine then told him of her last interview with Mrs. Reffold.

"Well, if you feel uncomfortable, it is as it should be," he said. "I don't see what business you had to point out to Mrs. Reffold her duty. I dare say she knows it quite well, though she may not choose to do it. I am sure I should resent it, if any one pointed out my duty to me. Every one knows his own duty. And it is his own affair whether or not he does it."

"I wonder if you are right," Bernardine said. "I never meant to presume; but her indifference had exasperated me."

"Why should you be exasperated about other people's affairs?" he said. "And why interfere at all?"

"Being interested is not the same as being interfering," she replied quickly.

"It is difficult to be the one without being the other," he said. "It requires a genius. There is a genius for being sympathetic as well as a genius for being good. And geniuses are few."

"But I knew one," Bernardine said. "There was a friend to whom in the first days of my trouble I turned for sympathy. When others only irritated, she could soothe. She had only to come into my room, and all was well with me."

There were tears in Bernardine's eyes as she spoke.

"Well," said the Disagreeable Man kindly, "and where is your genius now?"

"She went away, she and hers," Bernardine said. "And that was the end of that chapter."

"Poor little child," he said, half to himself.

"Don't I too know something about the ending of such a chapter?"

But Bernardine did not hear him; she was thinking of her friend. She was thinking, as we all think, that those to whom in our suffering we turn for sympathy, become hallowed beings. Saints they may not be; but for want of a better name, saints they are to us, gracious and lovely presences. The great time Eternity, the great space Death, could not rob them of their saintship; for they were canonized by our bitterest tears.

She was roused from her reverie by the Disagreeable Man, who got up, and pushed his chair noisily under the table.

"Will you come and help me to develop some photographs?" he asked cheerily. "You do not need to have a straight eye for that!"

Then as they went along together, he said:

"When we come to think about it seriously, it is rather absurd for us to expect to have uninterrupted stretches of happiness. Happiness falls to our share in separate detached bits; and those of us who are wise, content ourselves with these broken fragments."

"But who is wise?" Bernardine asked. "Why, we all expect to be happy. No one told us that we were to be happy. Still, though no one told us, it is the true instinct of human nature."

"It would be interesting to know at what particular period of evolution into our present glorious types we felt that instinct for the first time," he said. "The sunshine must have had something to do with it. You see how a dog throws itself down in the sunshine; the most wretched cur heaves a sigh of content then; the sulkiest cat begins to purr."

They were standing outside the room set apart for the photograph-maniacs of the Kurhaus.

"I cannot go into that horrid little hole," Bernardine said. "And besides, I have promised to play chess with the Swedish professor. And after that I am going to photograph Marie. I promised Wärli I would."

The Disagreeable Man smiled grimly.

"I hope he will be able to recognize her!" he said. Then, feeling that he was on dangerous ground, he added quickly:

"If you want any more plates, I can oblige you."

On her way to her room she stopped to talk to pretty Fräulein Müller, who was in high spirits, having had an excellent report from the Doctor. Fräulein Müller always insisted on talking English with Bernardine; and as her knowledge of it was limited, a certain amount of imagination was necessary to enable her to be understood.

"Ah, Miss Holme," she said, "I have deceived an exquisite report from the Doctor."

"You are looking ever so well," Bernardine said. "And the love-making with the Spanish gentleman goes on well, too?"

"Ach!" was the merry answer. "That is your inventory! I am quite indolent to him!"

At that moment the Spanish gentleman came out of the Kurhaus flower-shop, with a beautiful bouquet of flowers.

"Mademoiselle," he said, handing them to Fräulein Müller, and at the same time putting his hand to his heart. He had not noticed Bernardine at first, and when he saw her, he became somewhat confused. She smiled at them both, and escaped into the flower-shop, which was situated in one of the covered passages connecting the mother-building with the dependencies. Herr Schmidt, the gardener, was making a wreath. His favourite companion, a saffron cat, was playing with the wire. Schmidt was rather an ill-tempered man, but he liked Bernardine.

"I have put these violets aside for you, Fräulein," he said, in his sulky way. "I meant to have sent them to your room, but have been interrupted in my work."

"You spoil me with your gifts," she said.

"You spoil my cat with the milk," he replied, looking up from his work.

"That is a beautiful wreath you are making, Herr Schmidt," she said. "Who has died? Any one in the Kurhaus?"

"No, Fräulein. But I ought to keep my

door locked when I make these wreaths. People get frightened, and think they, too, are going to die. Shall you be frightened, I wonder?"

"No, I believe not," she answered as she took possession of her violets, and stroked the saffron cat. "But I am glad no one has died here."

"It is for a young, beautiful lady," he said. "She was in the Kurhaus two years ago. I liked her. So I am taking extra pains. She did not care for the flowers to be wired. So I am trying my best without the wire. But it is difficult."

She left him to his work, and went away, thinking. All the time she had now been in Petershof had not sufficed to make her indifferent to the sadness of her surroundings. In vain the Disagreeable Man's preachings, in vain her own reasonings with herself.

These people here who suffered, and faded, and passed away, who were they to her?

Why should the faintest shadow steal across her soul on account of them?

There was no reason. And still she felt for

them all, she who in the old days would have thought it waste of time to spare a moment's reflection on anything so unimportant as the sufferings of an *individual* human being.

And the bridge between her former and her present self was her own illness.

What dull-minded sheep we must all be, how lacking in the very elements of imagination, since we are only able to learn by personal experience of grief and suffering, something about the suffering and grief of others!

Yea, how the dogs must wonder at us: those dogs who know when we are in pain or trouble, and nestle nearer to us.

So Bernardine reached her own door. She heard her name called, and, turning round, saw Mrs. Reffold. There was a scared look on the beautiful face.

"Miss Holme," she said, "I have been sent for—I daren't go to him alone—I want you—he is worse. I am"....

Bernardine took her hand, and the two women hurried away in silence.

CHAPTER XVI.

WHEN THE SOUL KNOWS ITS OWN REMORSE.

BERNARDINE had seen Mr. Reffold the previous day. She had sat by his side and held his hand. He had smiled at her many times, but he only spoke once.

"Little Brick," he whispered—for his voice had become nothing but a whisper—"I remember all you told me. God bless you. But what a long time it *does* take to die."

But that was yesterday.

The lane had come to an ending at last, and Mr. Reffold lay dead.

They bore him to the little mortuary chapel. And Bernardine stayed with Mrs. Reffold, who seemed afraid to be alone. She clung to Bernardine's hand.

"No, no," she said excitedly, "you must not go! I can't bear to be alone: you must stay with me."

She expressed no sorrow, no regret. She did not even speak his name. She just sat nursing her beautiful face.

Once or twice Bernardine tried to slip away. This waiting about was a strain on her, and she felt that she was doing no good.

But each time Mrs. Reffold looked up and prevented her.

"No, no," she said. "I can't bear myself without you. I must have you near me. Why should you leave me?"

So Bernardine lingered. She tried to read a book which lay on the table. She counted the lines and dots on the wall-paper. She thought about the dead man; and about the living woman. She had pitied him; but when she looked at the stricken face of his wife, Bernardine's whole heart rose up in pity for her. Remorse would come, although it might not remain long. The soul would see itself face to face for one brief moment; and then forget its own likeness.

But for the moment—what a weight of suffering, what a whole century of agony!

Bernardine grew very tender for Mrs. Reffold: she bent over the sofa, and fondled the beautiful face.

"Mrs. Reffold" . . . she whispered.

That was all she said: but it was enough.

Mrs. Reffold burst into an agony of tears.

"Oh, Miss Holme," she sobbed, "and I was not even kind to him! And now it is too late. How can I ever bear myself?"

And then it was that the soul knew its own remorse.

CHAPTER XVII.

A RETURN TO OLD PASTURES.

She had left him alone and neglected for whole hours when he was alive. And now when he was dead, and it probably mattered little to him where he was laid, it was some time before she could make up her mind to leave him in the lonely little Petershof cemetery.

"It will be so dreary for him there," she said to the Doctor.

"Not so dreary as you made it for him here," thought the Doctor.

But he did not say that: he just urged her quietly to have her husband buried in Petershof; and she yielded.

So they laid him to rest in the dreary cemetery.

Bernardine went to the funeral, much against the Disagreeable Man's wish.

"You are looking like a ghost yourself," he said to her. "Come out with me into the country instead."

But she shook her head.

"Another day," she said. "And Mrs. Reffold wants me. I can't leave her alone, for she is so miserable."

The Disagreeable Man shrugged his shoulders, and went off by himself.

Mrs. Reffold clung very much to Bernardine those last days before she left Petershof. She had decided to go to Wiesbaden, where she had relations; and she invited Bernardine to go with her: it was more than that, she almost begged her. Bernardine refused.

"I have been from England nearly five months," she said, "and my money is coming to an end. I must go back and work."

"Then come away with me as my companion," Mrs. Reffold suggested. "And I will pay you a handsome salary."

Bernardine could not be persuaded.

"No," she said. "I could not earn money that way: it would not suit me. And besides, you would not care to be a long time with me: you would soon tire of me. You think you would like to have me with you now. But I know how it would be: you would be sorry, and so should I. So let us part as we are now: you going your way, and I going mine. We live in different worlds, Mrs. Reffold: it would be as senseless for me to venture into yours, as for you to come into mine. Do you think I am unkind?"

So they parted. Mrs. Reffold had spoken no word of affection to Bernardine, but at the station, as she bent down to kiss her, she whispered:

"I know you will not think too hardly of me. Still, will you promise me? And if you are ever in trouble, and I can help you, will you write to me?"

And Bernardine promised.

When she got back to her room, she found a small packet on her table. It contained Mr. Reffold's watch-chain. She had so often seen him playing with it. There was a little piece of paper enclosed with it, and Mr. Reffold had written on it some two months ago: "Give my watch-chain to Little Brick, if she will sacrifice a little of her pride, and accept the gift." Bernardine unfastened her watch from the black hair cord, and attached it instead to Mr. Reffold's massive gold chain.

As she sat there fiddling with it, the idea seized her that she would be all the better for a day's outing. At first she thought she would go alone, and then she decided to ask Robert Allitsen. She learnt from Marie that he was in the dark room, and she hastened down. She knocked several times before there was any answer.

"I can't be disturbed just now," he said. "Who is it?"

"I can't shout to you," she said.

The Disagreeable Man opened the door of the dark room.

"My negatives will be spoilt," he said gruffly. Then seeing Bernardine standing there, he added:

"Why, you look as though you wanted some brandy."

"No," she said, smiling at his sudden change of manner. "I want fresh air, a sledge drive, and a day's outing. Will you come?"

He made no answer, and retired once more into the dark room. Then he came out with his camera.

"We will go to that inn again," he said cheerily. "I want to take the photographs to those peasants."

In half an hour's time they were on their way. It was the same drive as before: and since then, Bernardine had seen more of the country, and was more accustomed to the wonderful white scenery: but still the "white presences" awed her, and still the deep silence held her. It was the same scene, and yet not the same either, for the season was now far advanced, and the melting of the snows had begun. In the far distance the whiteness seemed as before; but on the slopes near at hand, the green was beginning to assert itself, and some of the great trees had cast off their heavy burdens, and appeared more gloomy in their freedom than in the days of their snow-bondage. The roads were no longer quite so even as before; the sledge glided along when it could, and bumped along when it must. Still, there was sufficient snow left to make the drive possible, and even pleasant.

The two companions were quiet. Once only the Disagreeable Man made a remark, and then he said:

"I am afraid my negatives will be spoilt."

"You said that before," Bernardine remarked.

"Well, I say it again," he answered, in his grim way.

Then came a long pause.

"The best part of the winter is over," he said. "We may have some more snow; but it is more probable that we shall not. It is not enjoyable being here during the melting time."

"Well, in any case I should not be here much longer," she said; "and for a simple reason, too. I have nearly come to the end of my money. I shall have to go back and set to work again. I should not have been able to give myself this chance, but that my uncle spared me some of his money, to which I added my savings."

"Are you badly off?" the Disagreeable Man asked rather timidly.

"I have very few wants," she answered brightly. "And wealth is only a relative word, after all."

"It is a pity that you should go back to work so soon," he said half to himself. "You are only just better; and it is easy to lose what one has gained."

"Oh, I am not likely to lose," she answered; "but I shall be careful this time. I shall do a little teaching, and perhaps a little writing: not much—you need not be vexed. I shall not try to pick up the other threads yet. I shall not be political, nor educational, nor anything else great."

"If you call politics or education great," he said. "And heaven defend me from political or highly educated women!"

"You say that because you know nothing about them," she said sharply.

"Thank you," he replied. "I have met them quite often enough."

"That was probably some time ago," she said rather heartlessly. "If you have lived here so long, how can you judge of the changes which go on in the world outside Petershof?"

"If I have lived here so long," he repeated, in the bitterness of his heart.

Bernardine did not notice: she was on a subject which always excited her.

"I don't know so much about the political women," she said, "but I do know about the higher education people. The writers who rail against the women of this date are really describing the women of ten years ago. Why, the Girton girl of ten years ago seems a different creation from the Girton girl of to-day. Yet the latter has been the steady outgrowth of the former."

"And the difference between them?" asked the Disagreeable Man; "since you pride yourself on being so well informed."

"The Girton girl of ten years ago," said Bernardine, "was a sombre, spectacled person, carelessly and dowdily dressed, who gave herself up to wisdom and despised every one who did not know the Agamemnon by heart. She was probably not lovable; but she deserves to be honoured and thankfully remembered. She fought for woman's right to be well educated, and I cannot bear to hear her slighted. The fresh-hearted young girl who nowadays plays a good game of tennis, and takes a high place in the Classical or Mathematical Tripos, and is book learnèd, without being bookish, and . . . ."

"What other virtues are left, I wonder?" he interrupted.

"And who does not scorn to take a pride in her looks because she happens to take a pride in her books," continued Bernardine, looking at the Disagreeable Man, and not seeming to see him: "she is what she is by reason of that grave and loveless woman who won the battle for her."

Here she paused.

"But how ridiculous for me to talk to you in this way!" she said. "It is not likely that you would be interested in the widening out of women's lives."

"And pray why not?" he asked. "Have I been on the shelf too long?"

"I think you would not have been interested even if you had never been on the shelf," she said frankly. "You are not the type of man to be generous to woman."

"May I ask one little question of you, which shall conclude this subject," he said, "since here we are already at the Gasthaus: to which type of learned woman do you lay claim to belong?"

Bernardine laughed.

"That I leave to your own powers of discrimination," she said, and then added, "if you have any."

And that was the end of the matter, for the word spread about that Herr Allitsen had arrived, and every one turned out to give the two guests greeting. Frau Steinhart smothered Bernardine with motherly tenderness, and whispered in her ear:

"You are betrothed now, liebes Fräulein? Ach, I am sure of it."

But Bernardine smiled and shook her head, and went to greet the others who crowded round them; and at last poor Catharina drew near too, holding Bernardine's hand lovingly within her own. Then Hans, Liza's lover, came upon the scene, and Liza told the Disagreeable Man that she and Hans were to be married in a month's time. And the Disagreeable Man, much to Bernardine's amazement, drew from his pocket a small parcel, which he confided to Liza's care. Every one pressed round her while she opened it, and found what she had so often wished for, a silver watch and chain.

"Ach," she cried, "how heavenly! How all the girls here will envy me! How angry my dear friend Susanna will be!"

Then there were the photographs to be examined.

Liza looked with stubborn disapproval on the pictures of herself in her working-dress. But she did not conceal her admiration of the portraits which showed her to the world in her best finery.

"Ach," she cried, "this is something like a photograph!"

The Disagreeable Man grunted, but behaved after the fashion of a hero, claiming, however, a little silent sympathy from Bernardine.

It was a pleasant, homely scene: and Bernardine, who felt quite at her ease amongst these people, chatted away with them as though she had known them all her life.

Then Frau Steinhart suddenly remembered that her guests needed some food, and Liza was despatched to her duties as cook; though it was some time before she could be induced to leave off looking at the photographs.

"Take them with you, Liza," said the Disagreeable Man. "Then we shall get our meal all the quicker."

She ran off laughing, and finally Bernardine found herself alone with Catharina.

"Liza is very happy," she said to Bernardine. "She loves, and is loved."

"That is the greatest happiness," Bernardine said half to herself.

"Fräulein knows?" Catharina asked eagerly.

Bernardine looked wistfully at her companion.

"No, Catharina," she said. "I have only heard and read and seen."

"Then *you* cannot understand," Catharina said almost proudly. "But *I* understand."

She spoke no more after that, but took up her knitting, and watched Bernardine playing with the kittens. She was playing with the kittens, and she was thinking; and all the time she felt conscious that this peasant woman, stricken in mind and body, was pitying her because that great happiness of loving and being loved had not come into her life. It had seemed something apart from her; she had never even wanted it. She had wished to stand alone, like a little rock out at sea.

And now?

In a few minutes the Disagreeable Man and she sat down to their meal. In spite of her excitement, Liza managed to prepare everything nicely; though when she was making the omelette *aux fines herbes*, she had to be kept guarded lest she might run off to have another look at the silver watch and the photographs of herself in her finest frock!

Then Bernardine and Robert Allitsen drank to the health of Hans and Liza: and then came the time of reckoning. When he was paying the bill, Frau Steinhart, having given him the change, said coaxingly:

"Last time, you and Fräulein each paid a share: to-day you pay all. Then perhaps you are betrothed at last, dear Herr Allitsen? Ach, how the old Hausfrau wishes you happiness! Who deserves to be happy, if it is not our dear Herr Allitsen?"

"You have given me twenty centimes too much," he said quietly. "You have your head so full of other things that you cannot reckon properly."

But seeing that she looked troubled lest she might have offended him, he added quickly:

"When I am betrothed, good little old house-mother, you shall be the first to know."

And she had to be content with that. She asked no more questions of either of them: but she was terribly disappointed. There was something a little comical in her disappointment; but Robert Allitsen was not amused at it, as he had been on a former occasion. As he leaned back in the sledge, with the same girl for his companion, he recalled his feelings. He had been astonished and amused, and perhaps a little shy, and a great deal relieved that she had been sensible enough to be amused too.

And now?

They had been constantly together for many months: he who had never cared before for companionship, had found himself turning more and more to her.

And now he was going to lose her.

He looked up once or twice to make sure that she was still by his side: she sat there so quietly. At last he spoke in his usual gruff way.

"Have you exhausted all your eloquence in your oration about learned women?" he asked.

"No, I am reserving it for a better audience," she answered, trying to be bright. But she was not bright.

"I believe you came out to the country to-day to seek for cheerfulness," he said after a pause. " Have you found it ? "

" I do not know," she said. " It takes me some time to recover from shocks; and Mr. Reffold's death was a sorrow to me. What do you think about death ? Have you any theories about life and death, and the bridge between them ? Could you say anything to help one ? "

" Nothing," he answered. " Who could ? And by what means ? "

" Has there been no value in philosophy," she asked, "and the meditations of learnèd men ? "

" Philosophy ! " he sneered. " What has it done for us ? It has taught us some processes of the mind's working; taught us a few wonderful things which interest the few; but the centuries have come and gone, and the only thing which the whole human race pants to know, remains unknown: our beloved ones, shall we meet them, and how ?—the great secret of the universe. We ask for bread, and these philosophers give us a stone. What help could come from them : or from any one? Death is simply one of the hard facts of life."

" And the greatest evil," she said.

" We weave our romances about the next world," he continued ; " and any one who has a fresh romance to relate, or an old one dressed up in new language, will be listened to, and welcomed. That helps some people for a little while; and when the charm of the romance is over, then they are ready for another, perhaps more fantastic than the last. But the plot is always the same : our beloved ones—shall we meet them, and how? Isn't it pitiful? Why cannot we be more impersonal? These puny, petty minds of ours! When will they learn to expand ? "

" Why should we learn to be more impersonal ? " she said. " There was a time when I felt like that; but now I have learnt something better : that we need not be ashamed of being human ; above all, of having the best of human instincts, love, and the passionate wish for its continuance, and the unceasing grief at its withdrawal. There is no indignity in this;

nor any trace of weakmindedness in our restless craving to know about the Hereafter, and the possibilities of meeting again those whom we have lost here. It is right, and natural, and lovely that it should be the most important question. I know that many will say that there *are* weightier questions: they say so, but do they think so? Do we want to know first and foremost whether we shall do our work better elsewhere: whether we shall be endowed with more wisdom: whether, as poor Mr. Reffold said, we shall be glad to behave less like curs, and more like heroes? These questions come in, but they can be put aside. The other question can *never* be put on one side. If that were to become possible, it would only be so because the human heart had lost the best part of itself, its own humanity. We shall go on building our bridge between life and death, each one for himself. When we see that it is not strong enough, we shall break it down and build another. We shall watch other people building their bridges. We shall imitate, or criticise, or condemn. But as time goes on, we shall learn not to interfere, we shall know that one bridge is probably as good as the other; and that the greatest value of them all has been in the building of them. It does not matter what we build, but build we must: you, and I, and every one."

"I have long ceased to build my bridge," the Disagreeable Man said.

"It is an almost unconscious process," she said. "Perhaps you are still at work, or perhaps you are resting."

He shrugged his shoulders, and the two comrades fell into silence again.

They were within two miles of Petershof, when he broke the silence: there was something wonderfully gentle in his voice.

"You little thing," he said, "we are nearing home, and I have something to ask you. It is easier for me to ask here in the free open country, where the space seems to give us breathing room for our cramped lungs and minds."

"Well," she said kindly; she wondered what he could have to say.

"I am a little nervous of offending you," he continued, "and yet I trust you. It is only this. You said you had come to the end of your money, and that you must go home. It seems a pity when you are getting better. I have so much more than I need. I don't offer it to you as a gift, but I thought if you wished to stay longer, a loan from me would not be quite impossible to you. You could repay as quickly or as slowly as was convenient to you, and I should only be grateful and" . . . .

He stopped suddenly.

The tears had gathered in Bernardine's eyes; her hand rested for one moment on his arm.

"Mr. Allitsen," she said, "you did well to trust me. But I could not borrow money of any one, unless I was obliged. If I could of any one, it would have been of you. It is not a month ago since I was a little anxious about money; my remittances did not come. I thought then that if obliged to ask for temporary help, I should come to you: so you see if you have trusted me, I, too, have trusted you."

A smile passed over the Disagreeable Man's face, one of his rare, beautiful smiles.

"Supposing you change your mind," he said quietly, "you will not find that I have changed mine."

Then a few minutes brought them back to Petershof.

CHAPTER XVIII.

A BETROTHAL.

He had loved her so patiently, and now he felt that he must have his answer. It was only fair to her, and to himself too, that he should know exactly where he stood in her affections. She had certainly given him little signs here and there, which had made him believe that she was not indifferent to his admiration. Little signs were all very well for a short time; but meanwhile the season was coming to an end: she had told him that she was going back to her work at home. And then perhaps he would lose her altogether. It would not be safe now for him to delay a single day longer. So the little postman armed himself with courage.

Wärli's brain was muddled that day. He who prided himself upon knowing the names of all the guests in Petershof, made the most absurd mistakes about people and letters too; and received in acknowledgment of his stupidity a series of scoldings which would have unnerved a stronger person than the little hunchback postman.

In fact, he ceased to care how he gave out the letters: all the envelopes seemed to have the same name on them: *Marie Truog*. Every word which he tried to decipher turned to that; so finally he tried no more, leaving the destination of the letter to be decided by the impulse of the moment. At last he arrived at that quarter of the Kurhaus where Marie held sway. He heard her singing in her pantry. Suddenly she was summoned downstairs by an impatient bell-ringer, and on her return found Wärli waiting in the passage.

"What a goose you are!" she cried, throwing a letter at him; "you have left the wrong letter at No. 82."

Then some one else rang, and Marie hurried

off again. She came back with another letter in her hand, and found Würli sitting in her pantry.

"The wrong letter left at No. 54," she said, "and Madame in a horrid temper in consequence. What a nuisance you are to-day, Würli! Can't you read? Here, give the remaining letters to me. I'll sort them."

Würli took off his little round hat, and wiped his forehead.

"I can't read to-day, Marie," he said; "something has gone wrong with me. Every name I look at, turns to Marie Truog. I ought to have brought every one of the letters to you. But I knew they could not be all for you, though you have so many admirers. For they would not be likely to write at the same time, to catch the same post."

"It would be very dull if they did," said Marie, who was polishing some water-bottles with more diligence than was usual or even necessary.

"But I am the one who loves you, Mariechen," the little postman said. "I have always loved you ever since I can remember. I am not much to look at, Mariechen: the binding of the book is not beautiful, but the book itself is not a bad book."

Marie went on polishing the water-bottles. Then she held them up to the light to admire their unwonted cleanness.

"I don't plead for myself," continued Würli. "If you don't love me, that is the end of the matter. But if you do love me, Mariechen, and will marry me, you won't be unhappy. Now I have said all."

Marie put down the water-bottles, and turned to Würli.

"You have been a long time in telling me," she said, pouting. "Why didn't you tell me three months ago? It's too late now."

"Oh, Mariechen!" said the little postman, seizing her hand and covering it with kisses; "you love some one else—you are already betrothed? And now it's too late, and you love some one else!"

"I never said I loved some one else," Marie replied; "I only said it was too late. Why, it must be nearly five o'clock, and my lamps are not yet ready. I haven't a moment to spare. Dear me, and there is no oil in the can; no, not one little drop!"

"The devil take the oil!" exclaimed Würli, snatching the can out of her hands. "What do I want to know about the oil in the can? I want to know about the love in your heart. Oh, Mariechen, don't keep me waiting like this! Just tell me if you love me, and make me the merriest soul in all Switzerland."

"Must I tell the truth," she said, in a most melancholy tone of voice; "the truth and nothing else? Well, Würli, if you must know . . . . how I grieve to hurt you . . . ." Würli's heart sank, the tears came into his eyes. "But since it must be the truth, and nothing else," continued the torturer, ... "well, Fritz . . . . I love you!"

A few minutes afterwards, the Disagreeable Man, having failed to attract any notice by ringing, descended to Marie's pantry, to fetch his lamp. He discovered Würli embracing his betrothed,

"I am sorry to intrude," he said grimly, and he retreated at once. But directly afterwards he came back.

"The matron has just come upstairs," he said. And he hurried away.

CHAPTER XIX.

"SHIPS THAT SPEAK EACH OTHER IN PASSING.'

MANY of the guests in the foreign quarter had made a start downwards into the plains; and the Kurhaus itself, though still well filled with visitors, was every week losing some of its invalids. A few of the tables looked desolate, and some were not occupied at all, the lingerers having chosen, now that their party was broken up, to seek the refuge of another table. So that many stragglers found their way to the English dining-board, each bringing with him his own national bad manners, and causing much annoyance to the Disagreeable Man, who was a true John Bull in his contempt of all foreigners. The English table was, so he said, like England herself: the haven of other nation's offscourings.

There were several other signs, too, that the season was far advanced. The food had fallen off in quality and quantity. The invalids, some of them better and some of them worse, had become impatient. And plans were being discussed, where formerly temperatures and coughs and general symptoms were the usual subjects of conversation! The caretakers, too, were in a state of agitation; some few keenly anxious to be off to new pastures; and others, who had perhaps formed attachments, an occurrence not unusual in Petershof, were wishing to hold back time with both hands, and were therefore delighted that the weather, which had not yet broken up, gave no legitimate excuse for immediate departure.

Pretty Fräulein Müller had gone, leaving her Spanish gentleman quite disconsolate for the time being. The French Marchioness had returned to the Parisian circles where she was celebrated for all the domestic virtues, from which she had been taking such a prolonged holiday in Petershof. The little French danseuse and her poodle had left for Monte Carlo. M. Lichinsky and his mother passed on to the Tyrol, where Madame would no doubt have plenty of opportunities for quarrelling: or not finding them, would certainly make them without any delay, by this means keeping herself in good spirits and her son in bad health. There were some, too, who had hurried off without paying their doctors: being of course those who had received the greatest attention, and who had expressed the greatest gratitude in their time of trouble, but who were of opinion that thankfulness could very well take the place of francs: an opinion not entirely shared by the doctors themselves.

The Swedish professor had betaken himself off, with his chessmen and his chessboard. The little Polish governess who clutched so eagerly at her paltry winnings, caressing those centimes with the same fondness and fever that a greater gambler grasps his thousands of francs, she had left too; and, indeed, most of Bernardine's acquaintances had gone their several ways, after six months' constant intercourse and companionship, saying good-bye with the same indifference as though they were saying good-morning or good-afternoon.

This cold-heartedness struck Bernardine more than once, and she spoke of it to Robert Allitsen. It was the day before her own departure, and she had gone down with him to the restaurant, and sat sipping her coffee, and making her complaint.

"Such indifference is astonishing, and it is sad too. I cannot understand it," she said.

"That is because you are a goose," he replied, pouring out some more coffee for himself, and as an after thought, for her too. "You pretend to know something about the human heart, and yet you do not seem to grasp the fact that most of us are very little interested in other people: they for us and we for them can spare only a small fraction of time and attention. We may, perhaps, think to the contrary, believing that we occupy an important position in their lives; until one day, when we are feeling most confident of our value, we see an unmistakable sign, given quite unconsciously by our friends, that we are after all nothing to them: we can be done without, put on one side, and forgotten when not present. Then, if we are foolish, we are wounded by this discovery, and we draw back into ourselves. But if we are wise, we draw back into ourselves without being wounded: recognizing as fair and reasonable that people can only have time and attention for their immediate belongings. Isolated persons have to learn this lesson sooner or later; and the sooner they do learn it, the better."

"And you," she asked, "you have learnt this lesson?"

"Long ago," he said decidedly.

"You take a hard view of life," she said.

"Life has not been very bright for me," he answered. "But I own that I have not cultivated my garden. And now it is too late: the weeds have sprung up everywhere. Once or twice I have thought lately that I would begin to clear away the weeds, but I have not the courage now. And perhaps it does not matter much."

"I think it does matter," she said gently. "But I am no better than you, for I have not cultivated my garden."

"It would not be such a difficult business for you as for me," he said, smiling sadly.

They left the restaurant, and sauntered out together.

"And to-morrow you will be gone," he said.

"I shall miss you," Bernardine said.

"That is simply a question of time," he remarked. "I shall probably miss you at first. But we adjust ourselves easily to altered circumstances: mercifully. A few days, a few weeks at most, and then that state of becoming accustomed, called by pious folk, resignation."

"Then you think that the every-day companionship, the every-day exchange of thoughts and ideas, counts for little or nothing?" she asked.

"That is about the colour of it," he answered, in his old gruff way.

She thought of his words when she was

packing: the many pleasant hours were to count for nothing; for nothing the little bits of fun, the little displays of temper and vexation, the snatches of serious talk, the contradictions, and all the petty details of six months' close companionship.

He was not different from the others who had parted from her so lightly. No wonder, then, that he could sympathise with them.

That last night at Petershof, Bernardine hardened her heart against the Disagreeable Man.

"I am glad I am able to do so," she said to herself. "It makes it easier for me to go."

Then the vision of a forlorn figure rose before her. And the little hard heart softened at once.

In the morning they breakfasted together as usual. There was scarcely any conversation between them. He asked for her address, and she told him that she was going back to her uncle who kept the second-hand book-shop in Stone Street.

"I will send you a guide-book from the

Tyrol," he explained. "I shall be going there in a week or two to see my mother."

"I hope you will find her in good health," she said.

Then it suddenly flashed across her mind what he had told her about his one great sacrifice for his mother's sake. She looked up at him, and he met her glance without flinching.

He said good-bye to her at the foot of the staircase.

It was the first time she had ever shaken hands with him.

"Good-bye," he said gently. "Good luck to you."

"Good-bye," she answered.

He went up the stairs, and turned round as though he wished to say something more. But he changed his mind, and kept his own counsel.

An hour later Bernardine left Petershof. Only the concierge of the Kurhaus saw her off at the station.

CHAPTER XX.

A LOVE-LETTER.

Two days after Bernardine had left Petershof, the snows began to melt. Nothing could be drearier than that process: nothing more desolate than the outlook.

The Disagreeable Man sat in his bedroom trying to read Carpenter's Anatomy. It failed to hold him. Then he looked out of the window, and listened to the dripping of the icicles. At last he took a pen, and wrote as follows:

"LITTLE COMRADE, LITTLE PLAYMATE,

I could not believe that you were really going. When you first said that you would soon be leaving, I listened with unconcern, because it did not seem possible that the time could come when we should not be together; that the days would come and go, and that I should not know how you were; whether you were better, and more hopeful about your life and your work, or whether the old misery of indifference and ill-health was still clinging to you; whether your voice was strong as of one who had slept well and felt refreshed, or whether it was weak like that of one who had watched through the long night.

"It did not seem possible that such a time could come. Many cruel things have happened to me, as to scores of others, but this is the most cruel of all. Against my wish and against my knowledge, you have crept into my life as a necessity, and now I have to give you up. You are better, God bless you, and you go back to a fuller life, and to carry on your work, and to put to account those talents which no one realises more than I do; and as for myself, God help me, I am left to wither away.

"You little one, you dear little one, I never

wished to love you. I had never loved any one, never drawn near to any one. I have lived lonely all my young life; for I am only a young man yet. I said to myself time after time: 'I will not love her. It will not do me any good, nor her any good.' And then in my state of health, what right had I to think of marriage, and making a home for myself? Of course that was out of the question. And then I thought, that because I was a doomed man, cut off from the pleasures which make a lovely thing of life, it did not follow that I might not love you in my own quiet way, hugging my secret to myself, until the love became all the greater because it *was* my secret. I reasoned about it too: it could not harm you that I loved you. No one could be the worse for being loved. So little by little I yielded myself this luxury; and my heart once so dried up, began to flower again; yes, little one, you will smile when I tell you that my heart broke out into flower.

"When I think of it all now, I am not sorry that I let myself go. At least I have learnt what I know nothing of before: now I understand what people mean when they say that love adds a dignity to life which nothing else can give. That dignity is mine now, nothing can take it from me; it is my own. You are my very own; I love everything about you. From the beginning I recognized that you were clever and capable. Though I often made fun of what you said, that was simply a way I had; and when I saw you did not mind, I continued in that way, hoping always to vex you; your good temper provoked me, because I knew that you made allowances for me being a Petershof invalid. You would never have suffered a strong man to criticize you as I did; you would have flown at him, for you are a feverish little child: not a quiet woolly lamb. At first I was wild that you should make allowances for me. And then I gave in, as weak men are obliged. When you came, I saw that your troubles and sufferings would make you bitter. Do you know who helped to cure you? *It was I.* I have seen that often before. That is the one little bit of good I have done in the world: I have helped to cure cynicism. You were shocked at the things I said, and you were saved. I did not save you intentionally, so I am not posing as a philanthropist. I merely mention that you came here hard, and you went back tender. That was partly because you have lived in the City of Suffering. Some people live there and learn nothing. But you would learn to feel only too much. I wish that your capacity for feeling were less; but then you would not be yourself, your present self I mean, for you have changed even since I have known you. Every week you seemed to become more gentle. You thought me rough and gruff at parting, little comrade: I meant to be so. If you had only known, there was a whole world of tenderness for you in my heart. I could not trust myself to be tender to you; you would have guessed my secret. And I wanted you to go away undisturbed. You do not feel things lightly, and it was best for you that you should harden your heart against me.

"If you could harden your heart against me. But I am not sure about that. I believe that . . . .

Ah, well, I'm a foolish fellow; but some day, dear, I'll tell you what I think. . . . . I have treasured many of your sayings in my memory. I can never be as though I had never known you. Many of your words I have repeated to myself afterwards until they seemed to represent my own thoughts. I specially remember what you said about God having made us lonely, so that we might be obliged to turn to him. For we are all lonely, though some of us not quite so much as others. You yourself spoke often of being lonely. Oh, my own little one! Your loneliness is nothing compared to mine. How often I could have told you that.

"I have never seen any of your work, but I think you have now something to say to others, and that you will say it well. And if you have the courage to be simple when it comes to the point, you will succeed. And I believe you will have the courage, I believe everything of you.

"But whatever you do or do not, you will always be the same to me: my own little one, my very own. I have been waiting all my life for you; and I have given you my heart entire. If you only knew that, you could not call yourself lonely any more. If any one was ever loved, it is you, dear heart.

"Do you remember how those peasants at the Gasthaus thought we were betrothed? I thought that might annoy you; and though I was relieved at the time, still, later on, I wished you had been annoyed. That would have shown that you were not indifferent. From that time my love for you grew apace. You must not mind me telling you so often; I must go on telling you. Just think, dear, this is the first love-letter I have ever written: and every word of love is a whole world of love. I shall never call my life a failure now. I may have failed in everything else, but not in loving. Oh, little one, it can't be that I am not to be with you, and not to have you for my own! And yet how can that be? It is not I who may hold you in my arms. Some strong man must love and wrap you round with tenderness and softness. You little independent child, in spite of all your wonderful views and theories, you will soon be glad to lean on some one for comfort and sympathy. And then perhaps that troubled little spirit of yours may find its rest. Would to God I were that strong man!

"But because I love you, my own little darling, I will not spoil your life. I won't ask you to give me even one thought. But if I believed that it were of any good to say a prayer, I should pray that you may soon find that strong man; for it is not well for any of us to stand alone. There comes a time when the loneliness is more than we can bear.

"There is one thing I want you to know: indeed I am not the gruff fellow I have so often seemed. Do believe that. Do you remember how I told you that I dreamed of losing you? And now the dream has come true. I am always looking for you, and cannot find you.

"You have been very good to me; so patient, and genial, and frank. No one before has ever been so good. Even if I did not love you, I should say that.

"But I do love you, no one can take that from me: it is my own dignity, the crown of my life. Such a poor life .... no, no, I won't say that now. I cannot pity myself now .... no, I cannot. . . . "

The Disagreeable Man stopped writing, and the pen dropped on the table.

He buried his tear-stained face in his hands. He cried his heart out, this Disagreeable Man.

Then he took the letter which he had just been writing, and he tore it into fragments.

END OF PART I.

PART II.

CHAPTER I.

THE DUSTING OF THE BOOKS.

It was now more than three weeks since Bernardine's return to London. She had gone back to her old home, at her uncle's second-hand book-shop. She spent her time in dusting the books, and arranging them in some kind of order; for old Zerviah Holme had ceased to interest himself much in his belongings, and sat in the little inner room reading as usual Gibbon's "History of Rome." Customers might please themselves about coming: Zerviah Holme had never cared about amassing money, and now he cared even less than before. A frugal breakfast, a frugal dinner, a box full of snuff, and a shelf full of Gibbon were the old man's only requirements : an undemanding life, and therefore a loveless one; since the less we ask for, the less we get.

When Malvina his wife died, people said: "He will miss her."

But he did not seem to miss her : he took his breakfast, his pinch of snuff, his Gibbon, in precisely the same way as before, and in the same quantities.

When Bernardine first fell ill, people said: "He will be sorry. He is fond of her in his own queer way."

But he did not seem to be sorry. He did not understand anything about illness. The thought of it worried him; so he put it from him. He remembered vaguely that Bernardine's father had suddenly become ill, that his powers had all failed him, and that he lingered on, just a wreck of humanity, and then died. That was twenty years ago. Then he thought of Bernardine, and said to himself, "History repeats itself." That was all.

Unkind? No; for when it was told him

that she must go away, he looked at her wonderingly, and then went out. It was very rarely that he went out. He came back with fifty pounds.

"When that is done," he told her, "I can find more."

When she went away, people said: "He will be lonely."

But he did not seem to be lonely. They asked him once, and he said: "I always have Gibbon."

And when she came back, they said: "He will be glad."

But her return seemed to make no difference to him.

He looked at her in his usual sightless manner, and asked her what she intended to do.

"I shall dust the books," she said.

"Ah, I dare say they want it," he remarked.

"I shall get a little teaching to do," she continued. "And I shall take care of you."

"Ah," he said vaguely. He did not understand what she meant. She had never been very near to him, and he had never been very near to her. He had taken but little notice of her comings and goings; she had either never tried to win his interest or had failed: probably the latter. Now she was going to take care of him.

This was the home to which Bernardine had returned. She came back with many resolutions to help to make his old age bright. She looked back now, and saw how little she had given of herself to her aunt and her uncle. Aunt Malvina was dead, and Bernardine did not regret her. Uncle Zerviah was here still; she would be tender with him, and win his affection. She thought she could not begin better than by looking after his books. Each one was dusted carefully. The dingy old shop was restored to cleanliness. Bernardine became interested in her task. "I will work up the business," she thought. She did not care in the least about the books; she never looked into them except to clean them; but she was thankful to have the occupation at hand: something to help her over a difficult time. For the most trying part of an illness is when we are ill no longer; when there is no excuse for being idle and listless; when, in fact, we could work if we would: then is the moment for us to begin on anything which presents itself, until we have the courage and the inclination to go back to our own particular work: that which we have longed to do, and about which we now care nothing.

So Bernardine dusted books, and sometimes sold them. All the time she thought of the Disagreeable Man. She missed him in her life. She had never loved before, and she loved him. The forlorn figure rose before her, and her eyes filled with tears. Sometimes the tears fell on the books, and spotted them.

Still, on the whole she was bright; but she found things difficult. She had lost her old enthusiasms, and nothing yet had taken their place. She went back to the circle of her acquaintances, and found that she had slipped away from touch with them. Whilst she had been ill, they had been busily at work on matters social and educational and political.

P

She thought them hard, the women especially: they thought her weak. They were disappointed in her; she was now looking for the more human qualities in them, and she, too, was disappointed.

"You have changed," they said to her: "but then of course you have been ill, haven't you?"

With these strong, active people, to be ill and useless is a reproach. And Bernardine felt it as such. But she had changed, and she herself perceived it in many ways. It was not that she was necessarily better, but that she was different; probably more human, and probably less self-confident. She had lived in a world of books, and she had burst through that bondage and come out into a wider and a freer land.

New sorts of interests came into her life. What she had lost in strength, she had gained in tenderness. Her very manner was gentler, her mode of speech less assertive. At least, this was the criticism of those who had liked her but little before her illness.

"She has learnt," they said amongst themselves. And they were not scholars. They *knew*.

These, two or three of them, drew her nearer to them. She was alone there with the old man, and, though better, needed care. They mothered her as well as they could, at first timidly, and then with that sweet despotism which is for us all an easy yoke to bear. They were drawn to her as they had never been drawn before. They felt that she was no longer analysing them, weighing them in her intellectual balance, and finding them wanting; so they were free with her now, and revealed to her qualities at which she had never guessed before.

As the days went on, Zerviah began to notice that things were somehow different. He found some flowers near his table. He was reading about Nero at the time; but he put aside his Gibbon, and fondled the flowers instead. Bernardine did not know that.

One morning when she was out, he went into the shop and saw a great change there. Some one had been busy at work. The old man was pleased: he loved his books, though of late he had neglected them.

"She never used to take any interest in them," he said to himself. "I wonder why she does now?"

He began to count upon seeing her. When she came back from her outings, he was glad. But she did not know. If he had given any sign of welcome to her during those first difficult days, it would have been a great encouragement to her.

He watched her feeding the sparrows. One day when she was not there, he went and did the same. Another day when she had forgotten, he surprised her by reminding her.

"You have forgotten to feed the sparrows," he said. "They must be quite hungry."

That seemed to break the ice a little. The next morning when she was arranging some books in the old shop, he came in and watched her.

"It is a comfort to have you," he said. That was all he said, but Bernardine flushed with pleasure.

"I wish I had been more to you all these years," she said gently.

He did not quite take that in: and returned hastily to Gibbon.

Then they began to stroll out together. They had nothing to talk about: he was not interested in the outside world, and she was not interested in Roman History. But they were trying to get nearer to each other: they had lived years together, but they had never advanced a step; now they were trying, she consciously, he unconsciously. But it was a slow process, and pathetic, as everything human is.

"If we could only find some subject which we both liked," Bernardine thought to herself. "That might knit us together."

Well, they found a subject; though, perhaps, it was an unlikely one. The cart-horses: those great, strong, patient toilers of the road attracted their attention, and after that no walk was without its pleasure or interest. The brewers' horses were the favourites, though there were others, too, which met with their approval. He began to know and recognize them. He was almost like a child in his new-found interest. On Whit Monday they both went to the cart-horse parade in Regent's Park. They talked about the enjoyment for days afterwards.

"Next year," he told her, "we must subscribe to the fund, even if we have to sell a book."

He did not like to sell his books: he parted with them painfully, as some people part with their illusions.

Bernardine bought a paper for herself every day; but one evening she came in without one. She had been seeing after some teaching, and had without any difficulty succeeded in getting some temporary light work at one of the high schools. She forgot to buy her newspaper.

The old man noticed this. He put on his shabby felt hat, and went down the street, and brought in a copy of the *Daily News*.

"I don't remember what you like, but will this do?" he asked.

He was quite proud of himself for showing her this attention, almost as proud as the Disagreeable Man, when he did something kind and thoughtful.

Bernardine thought of him, and the tears came into her eyes at once. When did she not think of him? Then she glanced at the front sheet, and in the death column her eye rested on his name: and she read that Robert Allitsen's mother had passed away. So the Disagreeable Man had won his freedom at last. His words echoed back to her :

"But I know how to wait : if I have not learnt anything else, I have learnt how to wait. And some day I shall be free. And then . . . ."

CHAPTER II.

BERNARDINE BEGINS HER BOOK.

AFTER the announcement of Mrs. Allitsen's death, Bernardine lived in a misery of suspense. Every day she scanned the obituary, fearing to find the record of another death, fearing and yet wishing to know. The Disagreeable Man had yearned for his freedom these many years, and now he was at liberty to do what he chose with his poor life. It was of no value to him. Many a time she sat and shuddered. Many a time she began to write to him. Then she remembered that after all he had cared nothing for her companionship. He would not wish to hear from her. And besides, what had she to say to him?

A feeling of desolation came over her. It was not enough for her to take care of the old man who was drawing nearer to her every day; nor was it enough for her to dust the books, and serve any chance customers who might look in. In the midst of her trouble she remembered some of her old ambitions; and she turned to them for comfort as we turn to old friends.

"I will try to begin my book," she said to herself. "If I can only get interested in it, I shall forget my anxiety."

But the love of her work had left her. Bernardine fretted. She sat in the old bookshop, her pen unused, her paper uncovered. She was very miserable.

Then one evening when she was feeling that it was of no use trying to force herself to begin her book, she took her pen suddenly, and wrote the following prologue.

CHAPTER III.

FAILURE AND SUCCESS: A PROLOGUE.

Failure and Success passed away from Earth, and found themselves in a Foreign Land. Success still wore her laurel-wreath which she had won on Earth. There was a look of ease about her whole appearance; and there was a smile of pleasure and satisfaction on her face, as though she knew she had done well and had deserved her honours.

Failure's head was bowed: no laurel-wreath encircled it. Her face was wan, and pain-engraven. She had once been beautiful and hopeful, but she had long since lost both hope and beauty. They stood together, these two, waiting for an audience with the Sovereign of the Foreign Land. An old grey-haired man came to them and asked their names.

"I am Success," said Success, advancing a step forward, and smiling at him, and pointing to her laurel-wreath.

He shook his head.

"Ah," he said, "do not be too confident. Very often things go by opposites in this land. What you call Success, we often call Failure; what you call Failure, we call Success. Do you see those two men waiting there? The one nearer to us was thought to be a good man in your world; the other was generally accounted bad. But here we call the bad man good, and the good man bad. That seems strange to you. Well, then, look yonder. You considered that statesman to be sincere; but we say he was insincere. We chose as our poet-laureate a man at whom your world scoffed. Ay, and those flowers yonder: for us they have a fragrant charm; we love to see them near us. But you do not even take the trouble to pluck them from the hedges where they grow in rich profusion. So, you see, what we value as a treasure, you do not value at all."

Then he turned to Failure.

"And your name?" he asked kindly, though indeed he must have known it.

"I am Failure," she said sadly.

He took her by the hand.

"Come, now, Success," he said to her: "let me lead you into the Presence-Chamber."

Then she who had been called Failure, and was now called Success, lifted up her bowed head, and raised her weary frame, and smiled at the music of her new name. And with that smile she regained her beauty and her hope. And hope having come back to her, all her strength returned.

"But what of her?" she asked regretfully of the old grey-haired man; "must she be left?"

"She will learn," the old man whispered. "She is learning already. Come, now: we must not linger."

So she of the new name passed into the Presence-Chamber.

But the Sovereign said

"The world needs you, dear and honoured worker. You know your real name: do not heed what the world may call you. Go back and work, but take with you this time unconquerable hope."

So she went back and worked, taking with her unconquerable hope, and the sweet remembrance of the Sovereign's words, and the gracious music of her Real Name.

CHAPTER IV.

THE DISAGREEABLE MAN GIVES UP HIS FREEDOM.

THE morning after Bernardine began her book, she and old Zerviah were sitting together in the shop. He had come from the little inner room where he had been reading Gibbon for the last two hours. He still held the volume in his hand; but he did not continue reading, he watched her arranging the pages of a dilapidated book.

Suddenly she looked up from her work.

"Uncle Zerviah," she said brusquely, "you have lived through a long life, and must have passed through many different experiences. Was there ever a time when you cared for people rather than books?"

"Yes," he answered a little uneasily. He was not accustomed to have questions asked of him.

"Tell me about it," she said.

"It was long ago," he said half dreamily, "long before I married Malvina. And she died. That was all."

"That was all," repeated Bernardine, looking at him wonderingly. Then she drew nearer to him.

"And you have loved, Uncle Zerviah? And you were loved?"

"Yes, indeed," he answered, softly.

"Then you would not laugh at me if I were to unburden my heart to you?"

For answer, she felt the touch of his old hand on her head. And thus encouraged, she told him the story of the Disagreeable Man. She told him how she had never before loved any one until she loved the Disagreeable Man.

It was all very quietly told, in a simple and dignified manner: nevertheless, for all that, it was an unburdening of her heart; her listener being an old scholar who had almost forgotten the very name of love.

She was still talking, and he was still listening, when the shop door creaked. Zerviah crept quietly away, and Bernardine looked up.

The Disagreeable Man stood at the counter.

"You little thing," he said, "I have come to see you. It is eight years since I was in England."

Bernardine leaned over the counter.

"And you ought not to be here now," she said, looking at his thin face. He seemed to have shrunk away since she had last seen him.

"I am free to do what I choose," he said. "My mother is dead."

"I know," Bernardine said gently. "But you are not free."

He made no answer to that, but slipped into the chair.

"You look tired," he said. "What have you been doing?"

"I have been dusting the books," she answered, smiling at him. "You remember you told me I should be content to do that.

The very oldest and shabbiest have had my tenderest care. I found the shop in disorder. You see it now."

"I should not call it particularly tidy now," he said grimly. "Still, I suppose you have done your best. Well, and what else?"

"I have been trying to take care of my old uncle," she said. "We are just beginning to understand each other a little. And he is beginning to feel glad to have me. When I first discovered that, the days became easier to me. It makes us into dignified persons when we find out that there is a place for us to fill."

"Some people never find it out," he said.

"Probably, like myself, they went on for a long time, without caring," she answered. "I think I have had more luck than I deserve."

"Well," said the Disagreeable Man. "And you are glad to take up your life again?"

"No," she said quietly. "I have not got as far as that yet. But I believe that after some little time I may be glad: I hope so, I am working for that. Sometimes I begin to have a keen interest in everything. I wake up with an enthusiasm. After about two hours I have lost it again."

"Poor little child," he said tenderly. "I, too, know what that is. "But you *will* get back to gladness: not the same kind of satisfaction as before; but some other satisfaction, that compensation which is said to be included in the scheme."

"And I have begun my book," she said, pointing to a few sheets lying on the counter that is to say, I have written the Prologue."

"Then the dusting of the books has not sufficed?" he said, scanning her curiously.

"I wanted not to think of myself," Bernardine said. "Now that I have begun it, I shall enjoy going on with it. I hope it will be a companion to me."

"I wonder whether you will make a failure or a success of it?" he remarked. "I wish I could have seen."

"So you will," she said. "I shall finish it, and you will read it in Petershof."

"I shall not be going back to Petershof," he said. "Why should I go there now?"

"For the same reason that you went there eight years ago," she said.

"I went there for my mother's sake," he said.

"Then you will go there now for my sake," she said deliberately.

He looked up quickly.

"Little Bernardine," he cried, "my little Bernardine—is it possible that you care what becomes of me?"

She had been leaning against the counter, and now she raised herself, and stood erect, a proud, dignified little figure.

"Yes, I do care," she said simply, and with true earnestness. "I care with all my heart. And even if I did not care, you know, you would not be free. No one is free. You know that better than I do. We do not belong to ourselves: there are countless people depending on us, people whom we have never seen, and whom we never shall see. What we do, decides what they will be."

He still did not speak.

"But it is not for those others that I plead,"

she continued. "I plead for myself. I can't spare you, indeed, indeed I can't spare you! . . . ."

Her voice trembled, but she went on bravely:

"So you will go back to the mountains," she said. "You will live out your life like a man. Others may prove themselves cowards, but the Disagreeable Man has a better part to play."

He still did not speak. Was it that he could not trust himself to words? But in that brief time, the thoughts which passed through his mind were such as to overwhelm him. A picture rose up before him: a picture of a man and woman leading their lives together, each happy in the other's love; not a love born of fancy, but a love based on comradeship and true understanding of the soul. The picture faded, and the Disagreeable Man raised his eyes and looked at the little figure standing near him.

"Little child, little child," he said wearily, "since it is your wish, I will go back to the mountains."

Then he bent over the counter, and put his hand on hers.

"I will come and see you to-morrow," he said. "I think there are one or two things I want to say to you."

The next moment he was gone.

In the afternoon of that same day Bernardine went to the City. She was not unhappy: she had been making plans for herself. She would work hard, and fill her life as full as possible. There should be no room for unhealthy thought. She would go and spend her holidays in Petershof. There would be pleasure in that for him and for her. She would tell him so to-morrow. She knew he would be glad.

"Above all," she said to herself, "there shall be no room for unhealthy thought. I must cultivate my garden."

That was what she was thinking of at four in the afternoon: how she could best cultivate her garden.

At five she was lying unconscious in the accident-ward of the New Hospital: she had been knocked down by a waggon, and terribly injured.

"She will not recover," the Doctor said to

the nurse. "You see she is sinking rapidly. Poor little thing!"

At six she regained consciousness, and opened her eyes. The nurse bent over her. Then she whispered:

"Tell the Disagreeable Man how I wish I could have seen him to-morrow. We had so much to say to each other. And now . . . ."

The brown eyes looked at the nurse so entreatingly. It was a long time before she could forget the pathos of those brown eyes.

A few minutes later, she made another sign as though she wished to speak. Nurse Katharine bent nearer. Then she whispered:

"Tell the Disagreeable Man to go back to the mountains, and begin to build his bridge: it must be strong and . . . ."

Bernardine died.

CHAPTER V.

THE BUILDING OF THE BRIDGE.

ROBERT ALLITSEN came to the old book-shop to see Zerviah Holme before returning to the mountains. He found him reading Gibbon. These two men had stood by Bernardine's grave.

"I was beginning to know her," the old man said.

"I have always known her," the young man said. "I cannot remember a time when she has not been part of my life."

"She loved you," Zerviah said. "She was telling me so the very morning when you came."

Then, with a tenderness which was almost

foreign to him, Zerviah told Robert Allitsen how Bernardine had opened her heart to him. She had never loved any one before: but she had loved the Disagreeable Man.

"I did not love him because I was sorry for him," she had said. "I loved him for himself."

Those were her very words.

"Thank you," said the Disagreeable Man. "And God bless you for telling me."

Then he added:

"There were some few loose sheets of paper on the counter. She had begun her book. May I have them?"

Zerviah placed them in his hand.

"And this photograph," the old man said kindly. "I will spare it for you."

The picture of the little thin eager face was folded up with the papers.

The two men parted.

Zerviah Holme went back to his Roman History. The Disagreeable Man went back to the mountains: to live his life out there, and to build his bridge, as we all do, whether consciously or unconsciously. If it breaks down, we build it again.

"We will build it stronger this time," we say to ourselves.

So we begin once more.

We are very patient.

And meanwhile the years pass.

THE END.

16, Henrietta Street, Covent Garden, London.
October, 1893.

ALPHABETICAL LIST

OF

LAWRENCE & BULLEN'S PUBLICATIONS.

ALLEN, GRANT.— SCIENCE IN ARCADY. Crown 8vo. 5s.

"Holiday papers of a naturalist. The love of the country is in them all."—*Speaker.*

ANACREON.—The Greek Text, with THOMAS STANLEY'S Translation. Edited by A. H. BULLEN. Illustrated by J. R. WEGUELIN. Fcap. 4to. £1 1s. *net.*

ANDERSEN, HANS CHRISTIAN. — THE LITTLE MERMAID, AND OTHER STORIES. Translated by R. NISBET BAIN. With 65 Illustrations (chiefly full-page) by J. R. WEGUELIN. Royal 4to. 12s. 6d.

* Also 150 copies on hand-made paper, with the illustrations mounted on Japanese paper.

BARRETT, C. R. B.—ESSEX: HIGHWAYS, BYWAYS, AND WATERWAYS. First and Second Series. Written and Illustrated by C. R. B. BARRETT. (With 18 full-page etchings, and upwards of 200 drawings.) 2 vols. 12s. 6d. *net* per volume.

* 120 copies on fine paper, with additional etchings. Price £1 11s. 6d. *net* per volume.

"An excellent and original work."—*Athenæum.*

BARRETT, C. R. B.—THE TRINITY HOUSE OF DEPTFORD STROND. Royal 4to. 12s. 6d. *net*.

BARRETT, C. R. B.—ILLUSTRATED GUIDES.
1. Southwold. 6*d*.   2. Aldeburgh. 6*d*.
3. St. Osyth, Wivenhoe, Fingringhoe, and Brightlingsea. 6*d*.
4. Southend, Hadleigh, Rochford, &c. 6*d*.
5. Ipswich, Harwich, &c. 6*d*.
6. Great Yarmouth, 6*d*.   7. Caister Castle, 3*d*.
8. St. Osyth Priory, 3*d*.   9. Colchester, 6*d*.

"Carefully written, well printed, and amply illustrated.—*Manchester Guardian*.

BECKFORD, WILLIAM. — VATHEK. Edited by Dr. Richard Garnett. With 8 full-page Etchings by Herbert Nye. Demy 8vo. £1 1s. *net*.

* 600 copies printed for England and America. Also 70 copies on Japanese vellum, with an additional etching.

BOCCACCIO, GIOVANNI. — THE DECAMERON. Translated by John Payne. Illustrated by Louis Chalon. 2 vols. Imp. 8vo. £3 3s. *net*. (With 20 full-page Illustrations.)

* 1,000 copies printed for England and America.

BULLEN, A. H.—ANTHOLOGIES.
Lyrics from Elizabethan Song-Books. Revised edition. Fcp. 8vo. 5*s*.
Lyrics from Elizabethan Dramatists. Revised edition. Fcp. 8vo. 5*s*.

BULLEN, A. H.—ANTIENT DROLLERIES, in Six Parts. 3s. 6d. per Part *net*.

> * Parts I, II, and III. "Cobbe's Prophecies," "Pymlico, or Runne Redcap," and "Quips upon Questions," have appeared. Other Parts are in active preparation. The edition consists of 300 copies.

CATULLUS.—Edited by S. G. OWEN, Senior Student of Christ Church. Illustrated by J. R. WEGUELIN. Fcp. 4to. 16s. *net*.

> * Also 110 copies on Japanese vellum, with an additional illustration. Price £1 11s. 6d. *net*.

CHURCHILL, CHARLES.—ROSCIAD. Edited, with an Introduction and Notes, by ROBERT W. LOWE. With Portraits. Royal 4to. £1 1s' *net*.

> * The edition consists of 400 numbered copies.
>
> "The edition is not only good, but magnificent."—*Guardian*.

CHURCHILL, CHARLES. — PORTFOLIO OF PORTRAITS. 25s. *net*.

CRANE, WALTER.—CLAIMS OF DECORATIVE ART. Fcp. 4to. 7s. 6d. *net*.

> "No one has a better right than Mr. Walter Crane to write about the *Claims of Decorative Art*, for he is certainly one of the best masters of decorative design whom we have had among us for many a long day. . . . The book is admirably 'got up,' and does credit to the publishers."—*World*.

D'AULNOY, MADAME.—FAIRY TALES. Newly Translated into English, with an Introduction by ANNE THACKERAY RITCHIE, and Illustrations by CLINTON PETERS. Fcp. 4to. 7s. 6d.

"An admirable gift book for girls and boys."—*National Observer.*

"An exceedingly pleasing Volume."—*Saturday Review.*

* *Prospectus*, with specimen plate, on application.

DAVIDSON, JOHN. — SENTENCES AND PARAGRAPHS. 18mo. 3s. 6d.

EARLE, A. M.— CHINA - COLLECTING IN AMERICA. With Illustrations. Fcp. 4to. 16s.

EDMONDS, MRS.—THE HISTORY OF A CHURCH MOUSE. A modern Greek story. Crown 8vo. 1s. 6d.

"A graceful story, and one, moreover, which incidentally throws considerable light on the manners and customs of the Greek peasants in the more sequestered regions of that beautiful country at the present time."—*Speaker.*

GIFT, THEO.—FAIRY TALES FROM THE FAR EAST, Illustrated by O. VON GLEHN. Fcp. 4to. 5s.

'A charming volume adapted from the 'Birth Stories of Buddha,' as Englished by Professor Rhys Davies, with admirable drawings by Otto von Glehn."—*Saturday Review.*

GIFT, THEO.—AN ISLAND PRINCESS. A novel. 1 vol. Crown 8vo. 5s.

GISSING, GEORGE.—THE ODD WOMEN. A novel. 3 vols. 31s. 6d.

GISSING, GEORGE.—DENZIL QUARRIER. A novel. 1 vol. 6s.

GISSING, GEORGE.—THE EMANCIPATED. A novel. 1 vol. 6s. [*New and cheaper Edition.*

HARRADEN, BEATRICE.—SHIPS THAT PASS IN THE NIGHT. A novel. 1 vol. Crown 8vo. 3s. 6d. [*Fourth Edition.*

JÓKAI, MAURUS.—EYES LIKE THE SEA. A Romance. Translated from the Hungarian by R. NISBET BAIN. 3 vols. Crown 8vo. 31s. 6d.

KNIGHT, JOSEPH.—THEATRICAL NOTES (1874–1880). A contribution towards the History of the Modern English Stage. Demy 8vo. 6s.

* Also 250 large-paper copies, with portraits of eminent actors and actresses.

LINTON, W. J.—EUROPEAN REPUBLICANS. Recollections of Mazzini and his Friends. Demy 8vo. 10s. 6d.

"The book is one that cannot be read without some amount of searching of heart, for however our range of view, our political instincts have developed since '48, it would to-day be hard to find (save perhaps among the Russian and Polish exiles) so single-minded, unselfish, and devoted a band of politicians as these men, whom Mr. Linton revered in their lives and has fitly honoured after their death."—*Manchester Guardian.*

LINTON, W. J.—THE FLOWER AND THE STAR, and other Stories for Children. Written and Illustrated by W. J. LINTON. Fcp. 8vo. 3s. 6d.

" Delightfully fresh and unaffected. . . . The beautiful little woodcuts by the author form the most appropriate and expressive illustrations of such simple and pleasing stories that could be desired.—*Saturday Review.*

LINTON, W. J. — CATONINETALES. A Domestic Epic, by HATTIE BROWN, a young lady of colour lately deceased at the age of 14. Edited and Illustrated by W. J. LINTON. Demy 8vo. 7s. 6d. *net.* (330 copies printed.)

"The cat in the bag, on p. 48, though small, is too terrible."—*Saturday Review.*

MISCELLANIES—Bibliographical and Historical.

THE DIALOGUS OR COMMUNYNG BETWIXT THE WYSE KING SALOMON AND MARCOLPHUS. Reproduced in facsimile by the Oxford University Press from the unique copy of the edition printed by GERARD LEEU about 1492. Edited by E. GORDON

Duff. Small 4to. 10s. 6d. net. (350 copies printed.)

"Mr. Duff's edition possesses in a singular degree all the qualities which are necessary to justify a facsimile reprint."—*Guardian.*

ANTONIO DE GUARAS; OR, THE ACCESSION OF QUEEN MARY: being the Contemporary Narrative of Antonio de Guaras, a Spanish Merchant resident in London. Edited, with an Introduction, by RICHARD GARNETT, LL.D. Sm. 4to. 10s. 6d. net. (350 copies printed.)

"On the interest and importance of the narrative itself it is needless to dwell. . . . It is equally needless to say that Dr. Garnett has discharged his functions as editor in a masterly fashion."—*Times.*

SEX QUAM ELEGANTISSIME EPISTOLE IMPRESSE PER WILLELUM CAXTON ET DILIGENTER EMENDATE PER PETRUM CARMELIANUM. Reproduced in facsimile by JAMES HYATT. Edited, with a Translation, by GEORGE BULLEN, C.B., LL.D. Sm. 4to. 10s. 6d. net. (350 copies printed.)

"As a specimen of Caxtonian typography—and, we may add, of its artistic reproduction by means of photographic lithography—no less than on account of Dr. Bullen's exegetic labours, this reprint will be accounted curious and valuable."—*Times.*

INFORMACŌN FOR PYLGRYMES: Reproduced in facsimile by the Oxford University

Press from the unique copy preserved in the Advocates' Library at Edinburgh. Edited by E. GORDON DUFF. Sm. 4to. 10s. 6d. *net.* (350 copies printed.)

* A prospectus of the *Miscellanies* will be sent on application.

MUSES' LIBRARY—

POEMS OF WILLIAM BROWNE, OF TAVISTOCK. Edited by GORDON GOODWIN, with an Introduction by A. H. BULLEN. 2 vols 18mo. 10s. *net.*

POEMS OF WILLIAM BLAKE. Edited by W. B. YEATS. 1 vol. 18mo. 5s. *net.*

POEMS OF JOHN DONNE. Edited by E. K. CHAMBERS, with an Introduction by GEORGE SAINTSBURY. 2 vols. 18mo. 10s. *net.*

VOLUMES OF THE SERIES ALREADY ISSUED.

WORKS OF ROBERT HERRICK. Edited by A. W. POLLARD. With a Preface by A. C. SWINBURNE. 2 vols. 18mo. 10s. *net.*

POEMS AND SATIRES OF ANDREW MARVELL. Edited by G. A. AITKEN. 2 vols. 18mo. 10s. *net.*

POEMS OF EDMUND WALLER. Edited by G. THORN DRURY. 1 vol. 18mo. 5s. *net.*

POEMS OF JOHN GAY. Edited by J UNDERHILL. 2 vols. 18mo. 10s. *net.*

* Also 200 large-paper copies.
† Other volumes of the series are in active preparation.

O'NEILL, MOIRA.—AN EASTER VACATION. A story. Crown 8vo. 3s. 6d.

ORME, TEMPLE. — MATRICULATION CHEMISTRY. Small 8vo. 2s. 6d.

ORME, TEMPLE.—RUDIMENTS OF CHEMISTRY. Small 8vo. 2s. 6d.

OWEN, J. A. (Editor of "On Surrey Hills," &c.) **FOREST, FIELD AND FELL.** Crown 8vo. 3s. 6d.

PEARCE, J. H. (Author of "Esther Pentreath," &c.) **DROLLS FROM SHADOWLAND.** 18mo. 3s. 6d.

POWELL, G. H.—OCCASIONAL RHYMES AND REFLECTIONS. Demy 8vo. Boards, 1s. 6d. Cloth, 2s.

"Mr. Powell may fairly claim to share with Mr. Traill the laurels of modern English pasquinade."—*Times.*

PRIDEAUX, MISS S. T.—HISTORICAL SKETCH OF BOOKBINDING. (With a chapter "ON STAMPED BINDINGS," by E. GORDON DUFF.) Sm. 4to. 6s. net.

* Also 120 copies (numbered) on fine paper, with two facsimiles specially prepared by Mr. Griggs. £1 1s. net.

"We propose to consider the subject as it falls naturally into three main periods: the first from 1494, when Aldus Manutius had his printing press at Venice, to the end of the 16th century. This was the period of Maioli and Grolier, of the royal bindings done for Francis I. and Henri II. The art attained almost at once its highest perfection, at all events from the point of view of design. Secondly, the 17th century, with which are associated the names of the Eves and Le Gascon. Thirdly, the 18th century, the time of Boyat, Duseuil, Nicolas, and Antoine Padeloup and the Deromes, in France, and of the Harleian style and Roger Payne in England. Any division must necessarily be somewhat arbitrary, but it happens that in this case the centuries correspond pretty definitely to the different types of the art at different periods of its development."

RABELAIS, FRANCIS. — THE WORKS OF MASTER FRANCIS RABELAIS. Translated by Sir THOMAS URQUHART, of Cromarty, and PETER ANTONY MOTTEUX. With an Introduction by ANATOLE DE MONTAIGLON. Illustrated by L. CHALON. 2 vols. Imp. 8vo. £3 3s. net.

1,000 copies for England and America.

* *Prospectus*, with specimen plate, will be sent on application.

The copious racy vocabulary of Urquhart's "Rabelais," the odd quirks and flourishes, the gusto and swing of the rollicking narrative, can never fail to delight liberal readers.

The publishers of the present edition claim to have dealt handsomely with Rabelais and Sir Thomas Urquhart. They invited a very distinguished French artist, Mons. L. Chalon, to paint a series of oil-colour illustrations, which have been reproduced by Dujardin. The originals were lately exhibited at the "Blanc et Noir," Paris, where they were awarded a First Medal.

Prefixed to the translation is an essay on Rabelais (specially written for this edition) by a scholar of European reputation, M. Anatole de Montaiglon, whose knowledge of early French literature is certainly unsurpassed and probably unequalled. Facsimiles of rare title-pages of early French editions accompany the Introduction.

The volumes are printed by Messrs. Whittingham in the best style of the Chiswick Press.

ROBERTS, MORLEY.—KING BILLY OF BALLARAT, and other Tales. Crown 8vo. 5s.

"Mr. Roberts is a capital story-teller, with an incisive and dramatic style that is thoroughly individual.—*Saturday Review.*

ROBERTS, MORLEY.—SONGS OF ENERGY. Square 16mo. 5s.

ROBERTS, MORLEY.—LAND-TRAVEL AND SEA-FARING. With Illustrations by A. D. MCCORMICK. Demy 8vo. 7s. 6d.

ROBERTS, MORLEY.—THE MATE OF THE VANCOUVER. Crown 8vo. 3s. 6d.

ROBERTS, CECIL.—ADRIFT IN AMERICA; OR, WORK AND ADVENTURE IN THE STATES. Edited by MORLEY ROBERTS. Demy 8vo. 5s.

ROBINSON, H. J. — COLONIAL CHRONOLOGY: A chronology of the principal events connected with the English Colonies and India, from the close of the fifteenth century to the present time. With Maps. Crown folio. 16s.

"Nothing but cordial praise can be given to this valuable book."—*Manchester Guardian.*

"The book is one which ought to find a place in every library of reference."—*Speaker.*

"Admirably arranged on a plan equally simple and comprehensive."—*World.*

* *Prospectus* will be sent on application.

RUSSIAN FAIRY TALES.—Translated by R. NISBET BAIN. Illustrated by C. M. GERE Demy 8vo. 5s. [Second edition.

" The very best fairy-book that we have seen this year (or indeed for many years). . . . The six admirable full-page illustrations to 'Russian Fairy Tales,' by C. M. Gere (a name quite new to us by the way), approach as near to our ideal fairy-book pictures as may be. Messrs. Lawrence & Bullen are to be congratulated on having produced the most delightful story-book of the season."—*Daily Chronicle.*

"A book to read and a book to keep.—*Pall Mall Gazette.*

" Delightfully original, naïve and humorous."—*Truth.*

SCARRON, PAUL, COMICAL WORKS. Done into English by TOM BROWN of Shifnal. With an Introduction by J. J. JUSSERAND. Illustrated from the Designs of OUDRY. 2 vols. Demy 8vo. £1 1s. net.

* Also 150 copies on Japanese vellum. £2 2s. net.

" Published in a handsome form with every luxury of type and paper. A special feature consists in the designs by Oudry, the famous dog-painter to Louis XV. These are masterpieces of spirit and taste. . . . To the knowledge elsewhere accessible concerning the book, M. Jusserand now adds a brilliant account of the author."—*Athenæum.*

STRANG, WILLIAM.—**DEATH AND THE PLOUGHMAN'S WIFE.** A Ballad. With 9 Etchings and 2 Mezzotint Engravings. Folio.

* The price and the number of copies will be announced shortly.

TOLD IN THE VERANDAH.—Passages in the Life of Colonel Bowlong, set down by his Adjutant. Crown 8vo. 3s. 6d.

[*Third edition.*

"Colonel Bowlong is a liar of the first water. He recks not whether he deals with tiger-stories or with his alleged noble deeds on the field of battle. His tiger-story is one of the best we have ever read."—*St. James's Gazette.*

BY THE AUTHOR OF "TOLD IN THE VERANDAH." — A BLACK PRINCE AND OTHER STORIES. Crown 8vo. 3s. 6d.

VANBRUGH, SIR JOHN.—WORKS. Edited by W. C. WARD. 2 vols. Demy 8vo. (With a Portrait.) £1 5s. net.

WALLIS, HENRY.—PERSIAN AND ORIENTAL CERAMIC ART. Parts I. and II. Folio. 14s. net.

WELLS, CHARLES.—STORIES AFTER NATURE. With a Preface by W. J. LINTON. Fcp. 8vo. 7s. 6d. net.

* The edition consists of 400 numbered copies.

"The tales, with all their rouge and frippery of form, breathe a singularly clear and upright morality, and are rich in examples of noble manhood and gracious womanhood."—*Athenæum.*

WILLS, C. J.—JOHN SQUIRE'S SECRET. A novel. 1 vol. 3s. 6d. [*New and cheaper edition.*

YEATS, W. B.—THE CELTIC TWILIGHT. 18mo. 3s. 6d.

LIST OF PUBLICATIONS.

Books published at £3 3s. net.
BOCCACCIO'S *Decameron*. 2 vols.
URQUHART'S *Rabelais*. 2 vols.

3-Vol. Novels at £1 11s. 6d.
G. GISSING'S *Odd Women*.
JÓKAI'S *Eyes Like the Sea*.

£1 5s. net.
VANBRUGH'S *Plays*. 2 vols.
PORTRAITS ILLUSTRATING CHURCHILL'S *Rosciad*.

£1 1s. net.
ANACREON. | BECKFORD'S *Vathek*.
CHURCHILL'S *Rosciad*.
PAUL SCARRON.

16s. net.
CATULLUS.

16s.
ROBINSON'S *Colonial Chronology*.
EARLE'S *China Collecting*.

14s. net.
PARTS I. and II. OF WALLIS'S *Oriental Ceramic Art*.

12s. 6d. net.
BARRETT'S *Essex*.
BARRETT'S *Trinity House*.

12s. 6d.
HANS ANDERSEN'S *Little Mermaid*.

10s. 6d. net.
SALOMON AND MARCOLPHUS.
ANTONIO DE GUARAS.
SEX QUAM ELEGANTISSIMÆ EPISTOLE.
INFORMACŌN FOR PYLGRYMES.

10s. 6d.
LINTON'S *European Republicans*.

7s. 6d. net.
LINTON'S *Catoninetales*.
WELLS' *Stories After Nature*.
WALTER CRANE'S *Decorative Art*.

7s. 6d.
D'AULNOY'S *Fairy Tales*.
MORLEY ROBERTS' *Land Travel*.

6s. net.
MISS PRIDEAUX'S *Bookbinding*.

6s.
GISSING, G., *Denzil Quarrier*.
" " *The Emancipated*.
KNIGHT'S *Theatrical Notes*.

5s. net per Volume.
WILLIAM BROWNE, of Tavistock.
WILLIAM BLAKE. | JOHN DONNE.
JOHN GAY. | ROBERT HERRICK.
ANDREW MARVELL.
EDMUND WALLER.

5s.
GRANT ALLEN'S *Science in Arcady*.
BULLEN, A. H. *Lyrics from Song-Books*.
BULLEN, A. H. *Lyrics from Dramatists*.
GIFT, THEO. *Fairy Tales*.
" " *An Island Princess*.
ROBERTS, MORLEY. *King Billy*.
ROBERTS, C. *Adrift in America*.

3s. 6d. net per Part.
Antient Drolleries.

3s. 6d.
DAVIDSON'S *Sentences*.
HARRADEN'S *Ships That Pass*.
LINTON'S *Flower and the Star*.
O'NEILL'S *Easter Vacation*.
OWEN, J. A., *Woodland Ways*.
PEARCE'S *Drolls from Shadowland*.
ROBERTS' *Mate of Vancouver*.
"*Told in the Verandah*."
"*A Black Prince*."
WILLS' *John Squire's Secret*.
YEATS' *Celtic Twilight*.

2s. 6d.
ORME'S *Rudiments of Chemistry*.
" *Matriculation*.

1s. 6d.
EDMONDS, MRS., *Church Mouse*.
POWELL, G. H., *Rhymes*.

6d. and 3d.
BARRETT'S *Illustrated Guides*.

www.ingramcontent.com/pod-product-compliance
Lightning Source LLC
Chambersburg PA
CBHW021409230426
43666CB00006B/688